Neighborhoods and Crime

Neighborhoods and Crime

The Dimensions of Effective Community Control

Robert J. Bursik, Jr.
Harold G. Grasmick

LEXINGTON BOOKS
Lanham • Boulder • New York • Oxford

Published in the United States of America
by Lexington Books
4720 Boston Way, Lanham, Maryland 20706

12 Hid's Copse Road
Cumnor Hill, Oxford OX2 9JJ, England

Copyright © 1993 by Lexington Books
First paperback edition published 2001

British Library Cataloguing in Publication Information Available

The hardcover edition of this book was previously catalogued by the Library of Congress
as follows:

Bursik, Robert.
    Neighborhoods and crime: the dimensions of effective community control / Robert
J. Bursik, Jr., Harold G. Grasmick
        p.   cm.
    Includes index.
    ISBN 0-669-2462-8 (alk. paper)
        1. Crime—United States.   2. Neigborhoods—United States.   3. Social
Control—United States.   4. Crime prevention—United States.
I. Grasmick, Harold G. II. Title.
HV6789.B87 1993
364.4'3'0973—dc20   92-29041
                                                                            CIP

ISBN 0-7391-0302-4 (pbk. : alk. paper)

Printed in the United States of America

For Travis and Jake

Contents

Preface

Sociological theories offered as explanations of crime and delinquency have been characterized by an interesting historical cycle. The "classic" theories of the first part of the twentieth century, for example, had a distinct emphasis on the group aspects of such behavior. Therefore, neighborhoods (as well as peer groups in general and gangs in particular) were integral conceptual components of many of these early approaches. However, with the refinement of survey approaches to data collection and the increased interest in social-psychological theories of control, deterrence, learning, and labeling, the focus of the discipline significantly began to shift from group dynamics to individual processes during the 1960s and 1970s (see the discussions of Stark 1987; Bursik 1988; Krisberg 1991). While these new orientations provided centrally important insights into the highly complex phenomenon of crime, group-related criminological research became a decidedly secondary priority of the discipline.

However, the pendulum has begun to swing in the other direction, and there has been a relatively recent acceleration in the number of studies that have been conducted with an explicit focus on neighborhood dynamics. One of the hallmarks of this development has been an increasing focus on the neighborhood-based networks of association in which individuals are embedded and the implications of these relationships for the control of crime and delinquency. It is in this spirit that we have written this book.

The thesis that has been developed in this book has its roots in a series of studies conducted by Bursik concerning the rates of delinquency in Chicago neighborhoods. The initial studies were conducted at the Institute for Juvenile Research, which had sponsored the pioneering neighborhood research of Clifford Shaw and Henry McKay. Therefore, it seemed natural to ground those first efforts in Shaw and McKay's model of social disorganization. Although the degree to which a fifty-year-old theory could still account for variation in these rates was more than a little impressive, it was obvious that the dramatic changes in urban dynamics that had characterized the United States after World War II complicated the situation significantly.

This did not mean that Shaw and McKay were wrong; in fact, we are continually astonished by the richness of the theoretical insights presented by these criminologists, some of which have yet to be fully explored. In fact, as the material to be presented in this book will illustrate, the most fully developed aspects of their model, which focused on the internal dynamics of local communities and the capacity of local residents to regulate the behavior of their fellow neighbors, continue to be significantly related to neighborhood variations in crime rates. Nevertheless, although their ecological framework implies that urban areas are characterized by a system of interdependent communities, the role of these external relationships in the generation and control of crime and delinquency received very little attention from Shaw and McKay. This shortcoming makes the traditional social disorganization framework an incomplete representation of the dynamics that shape the regulatory capacities of contemporary urban neighborhoods that must compete with other local communities for scarce (and often shrinking) public and private resources.

In this book we argue that this shortcoming of the social disorganization perspective can be addressed by reformulating it within a broader systemic theory of community, which emphasizes how neighborhood life is shaped by the structure of formal and informal networks of association. Not only is the orientation of a systemic model consistent with social disorganization in its discussion of the regulatory capacities of networks embedded within the neighborhood, but it also formally addresses two aspects of community structure that Shaw and McKay virtually ignored: the networks among residents and local institutions, and the networks among local representatives of the neighborhood and external actors, institutions, and agencies. We feel that such a restatement and extension of the traditional social disorganization model makes it a powerful and viable framework for understanding the variable capacities of contemporary communities to control crime.

We certainly do not intend to claim that we have made any unique discovery concerning the systemic implications of the traditional social disorganization model, for others also have recognized the obvious connection between the two frameworks. Rather, we have been struck by the systemic implications of the processes of formal and informal control that have been analyzed in many areas of neighborhood criminology that traditionally have not been approached from a social disorganization perspective. Therefore, we felt that a systemic model of local community control could integrate effectively a broad range of criminological concerns at the neighborhood level. As a result, this book does not represent a work of primary data analysis per se but, rather, a synthesis of the contemporary body of literature within a single theoretical framework with important policy implications. Our exemplar in this goal has been Ruth Kornhauser's (1978) masterful explication of control theory. The degree to which we have convinced the reader that the systemic model provides a consistent, unifying framework for the analysis of these issues is, of course, a matter that only our readers can determine.

We have taken special care throughout this book to point out the measurement problems endemic to neighborhood-based crime research. Since direct indicators of many of the key processes discussed by major neighborhood theories of crime are not readily available to most researchers, many conclusions that have been drawn concerning the validity of particular neighborhood perspectives actually are based on analyses grounded in variables that represent very weak empirical realizations of complex processes. In fact, if we are truly honest with ourselves, the measurement of the central concepts underlying all theories of neighborhoods and crime (and not just systemic approaches) is still a fairly crude and indirect procedure. The key secondary goal of this book has been to highlight the degree to which these limitations have led to inconsistent results and have hindered the development and testing of sophisticated models of crime at the neighborhood level.

Chapter 1 develops the basic framework of our work and provides the reader with an introduction to our working conception of the neighborhood, systemic control, and crime. In each of the subsequent chapters, a particular aspect of these themes is explored in detail. Chapter 2 focuses on the most traditional question in the body of neighborhood research: why do the residents of some communities commit more crimes than residents of other communities? In this chapter we attempt to clear up some long-standing misunderstandings concerning the social disorganization model and restate the approach within a systemic framework. In addition, an important question is raised concerning the meaning of a neighborhood effect; that is, are the community differences that have been observed actually a function of the processes of systemic control, or do they simply present differences in the population composition of the area? As we will see, this is an important consideration when addressing the implications of differential racial and ethnic rates of criminal activity.

Chapter 3 reverses the focus of Chapter 2 to consider the question of why residents of particular neighborhoods tend to be victimized at higher rates than residents of other areas. We draw very heavily from the routine activities approach of Felson and Cohen (1980) to understand how systemic control is related to such victimization. In Chapter 4 we change our orientation from the neighborhood as the unit of analysis to the resident of that neighborhood to examine the systemic sources of fear of crime.

Chapter 5 addresses the implications of the systemic control approach for understanding an issue that has emerged as a key component of local and national concern about crime in urban neighborhoods: gangs and gang behavior. Chapter 6 examines various attempts that have been made by local communities to create and maintain neighborhood-based crime prevention programs. Policy implications are then discussed in light of these findings. Finally, we close the book with a short epilogue concerning the future of systemic approaches in criminology.

1

Basic Issues

> ... the festivities of prostitution, the orgies of pauperism, the haunts
> of theft and murder, the scenes of drunkenness and beastly debauch
> and all the sad realities that go to make up the lower stratum
> —the underground story of life in New York City.

George Foster's observations of a New York city neighborhood echo widely held sentiments concerning the pervasiveness of urban decline and decay in many of the major metropolitan areas of the United States. For example, John DiIulio (1989:39) has used similar imagery to describe a typical inner-city neighborhood as "a place where children walk to school through a maze of brazen drug dealers, and where owners of small shops keep guns beneath registers and close early on nights when they sense trouble. . . . [I]n short, it is a place hospitable to predatory street criminals."

Therefore, it may surprise some readers that Foster's comments were made in 1849, 150 years before those of DiIulio (see Ward 1989:17). In fact, Foster's statement is representative of many commentaries that were made concerning the quality of life in poor urban areas in the mid-nineteenth-century United States. For example, James Haskins (1974:24) has vividly described the Five Points–Paradise Square district of New York (which was located at the current intersection of Worth, Baxter, Park, and Mulberry Streets) as a "hell [where] pickpockets, murderers, thieves and all manner of thugs abounded" during the 1920s. Thus, despite occasional references to some mythical golden era of peace and tranquility in American neighborhoods, the current concern continues a long-standing tradition of journalistic and academic commentary, reflection, and intervention focused on the perceived problems of urban life.

This concern reflects much more than a simple interest in public safety and social welfare. For many Americans, the neighborhood symbolizes "that image that we call home" (Fried 1986:332–333) and serves as a haven of safety and belonging. This is especially true for young people, who generally have not had the opportunity to establish extensive networks of association outside of a relatively restricted geographic area. As Henry McKay (1949:32–33) observed, neighborhood institutions and activities are "the

1

most meaningful part of the child's social world." In addition, although many adults have networks of trans-neighborhood affiliations that can make their involvement in local communities much more limited than those of children (see, for example, Janowitz 1951; Fischer 1982), adult friendships still have a marked spatial concentration in many cities.[1] For example, between 36 and 41 percent of the residents of Detroit indicate that a majority of their friends live within the same neighborhood; less than one-third report that they have no friends within ten minutes of their homes (see Huckfeldt 1983:659–660).

There have been many significant changes in the traditional character and nature of American neighborhoods that have been generated by recent developments in the political, economic, and social dynamics of urban areas, such as the loss of employment opportunities in traditionally industrial cities, the increasing concentration of the extremely poor in the central city, the political disenfranchisement of many of those poor neighborhoods, and the erosion of local tax bases in some cities that has led to a decline in the abilty to provide essential public services. Nevertheless, the local neighborhood continues to provide an important frame of reference for the actions of its residents.

This process has been illustrated forcefully in Gerald Suttles' (1968) description of the "Addams Area," an older neighborhood of Chicago just southwest of the downtown (Loop) area. Although it is considered by the general public to be a primarily Italian community, it is populated by sizable populations of Italians, Mexicans, African-Americans, and Puerto Ricans. Owing to the history of the area, mutual distrust and suspicion is the primary characteristic of the relationships among the ethnicities residing in this area.

Nevertheless, this divided neighborhood was able to unite temporarily when a potentially threatening situation arose. In 1959 the University of Illinois announced plans to build a Chicago campus within the eastern portion of the neighborhood. A significant portion of the community was to be demolished for the construction of the campus, leading to the projected displacement of about one-third of the area's population. While local opposition to this plan eventually was unsuccessful, Suttles notes (p. 22) that members of all four groups joined forces in expressing concern over a shared threat, and at least one interethnic organization was formed to resist the construction.

Although he was not associated with the University, the first author of this book lived directly across the street from the Addams area campus from 1978 to 1983. Even after twenty years, a great deal of hostility and resentment continued to be directed toward the University and those connected with it. Partly in an effort to control the number of faculty, staff, and students who could move into the area, notifications of housing vacancies usually were made public only by word of mouth among long-time residents and were filled on the basis of personal recommendations. Likewise, despite the fact that the University was almost exclusively a commuter school, the residents in the area immediately surrounding the campus were successful in having their streets zoned for "Residential Parking Only," which made parking nearly impossible for those unable or unwilling to pay for the University facility. This antagonism was most notable in the obvious efforts to exclude the University from anything more than marginal participation in the vibrant social life found in the local taverns, restaurants, and businesses. Thus, although the war with the University had been lost, the battle continued.

Such conflicts over neighborhood control are a common feature of American urban life. Perhaps the most visible and volatile type of conflict has reflected the efforts of neighborhood residents to maintain the existing racial and ethnic composition of an area. Heitgerd and Bursik (1987), for example, have discussed a case in which the adult residents of a white ethnic community in Chicago fostered the use of delinquent behavior to discourage the movement of neighboring blacks into the neighborhood. A similar case in Philadelphia has been described by Wesley Skogan (1990:25).

Nearly as visible (and closely related; see Wilson 1991) are class-based conflicts over the maintenance of the socioeconomic composition of a neighborhood. Residents of relatively high-status areas may attempt to protect the reputation of their neighborhoods and the financial investments they have made in those areas by making it difficult for those with fewer economic resources to maintain homes in the community. The symbolic aspects of these conflicts can be quite amusing to outside observers, such as in 1985 when some residents of the very affluent Nichols Hills section of Oklahoma City succeeded in prohibiting the overnight parking of commercial and noncommercial pickup trucks on all streets and driveways within its boundaries.[2]

Although this pattern may represent the most common form of class-based neighborhood conflict, several cities also have experienced a sometimes violent resistance to the movement of upper-middle-class residents into traditionally lower- and working-class areas, that is, the process that has been referred to as gentrification. For example, the Knight-Ridder News Service (1991) released a story about the tensions that have evolved between the Polish, Italian, and German blue-collar residents of the Manayunk section of Philadelphia and the recent wave of artists, white-collar workers, and merchants who have tried to transform its working-class shopping district into a "hip row of pastel-colored shops that some compare to Coconut Grove without the palms or Sausalito without the San Francisco." The resistance to this change rapidly escalated from the exchange of verbal insults to broken windows, scratched cars, and assaults.[3]

These examples represent the efforts of local residents to regulate the nature of the activities that take place within the borders of their local communities. Throughout this book, we assume that the capacity for such regulation is determined by the extensiveness and density of the formal and informal networks within the neighborhood that bind the residents together as a social community. Owing to this emphasis on relational networks, our approach to neighborhood control is firmly grounded in the systemic theory of community organization (Berry and Kasarda 1977). In particular, we will argue that the differential rates of criminal behavior and victimization among neighborhoods, and the resulting fear of crime that may develop among the residents of crime-ridden areas, represent variations in the ability of neighborhoods to regulate themselves through these networks in such a way that the daily lives of their residents are not significantly constrained by the threat of criminal behavior. The success of such efforts is central to the perception of the quality of life within an area and provides the key context for the interpretation and evaluation of all other activities within that area.

Crime represents a very pervasive threat to neighborhood life. In the cases of racial/ethnic and class-based neighborhood conflicts that have been described, a community has tried to control the likelihood of a perceived undesirable change in its internal composition due to the influx of population groups located primarily outside of its boundaries. The situation is much more complicated in the control of crime, for the sources of threat to neighborhood life are internal as well as external. Alicia Rand (1986), for example, finds that almost 31 percent of all the crimes ever committed by youths born in Philadelphia during 1958 took place in the same census tract in which the youths resided; this ranged from a high of 51 percent for rapes and homicides to a low of 15 percent for larcenies.[4]

This creates some special difficulties in understanding attempts to control crime at the neighborhood level. For example, much of this book will discuss traditional approaches to the study of neighborhoods and crime that have focused on the characteristics and dynamics that are assumed to decrease the ability of a neighborhood to regulate itself, thereby increasing the likelihood that its residents will either engage in crime or become the victims of such activity. Although this work has provided some important findings concerning the relationship between systemic control and crime, the results of Bursik (1986b) and the discussion of Skogan (1990) suggest that it is not a simple linear process. Rather, high rates of crime can also significantly disrupt the capacity for maintaining control in the neighborhood. For example, residents may withdraw from participation in community affairs because of their heightened fear and anxiety. If such withdrawal from local networks becomes widespread, the sense of mutual responsibility among the residents is undermined, and those who are able to do so may attempt to physically abandon the neighborhood at the earliest possibility (Skogan 1990:13). As a result, the capacity for local control may further deteriorate, thereby accelerating the processes that originally gave rise to crime. In this book, we will pay special atention to the interaction between crime and the capacity for neighborhood control.

It has also been traditional to examine the neighborhood/crime relationship without a consideration of how the characteristics and dynamics of a particular area may be sensitive to the events occurring in other areas located in the same urban system or may be shaped by the economic, political, and social context of the broader metropolitan area. However, as Harvey Molotch (1976:310–311) argues, cities represent a system of competing neighborhood-based land interests that are capable of strategic coalition and action vis-à-vis other neighborhoods in that system (see also Logan and Molotch 1987). Skogan (1990:22) describes a striking example of such interest-based action in which the residents of a Chicago neighborhood put strong pressure on the Park District *not* to construct a park in their area for fear of the problems that might arise.

The consideration of such economic and political issues is essential for a full understanding of the relationship between neighborhoods and crime, for as Wacquant and Wilson have argued (1989:13–15), the increasing exclusion of blacks from the urban labor market has resulted in a growing isolation of residents in primarily black neighborhoods and a loss of neighborhood "organizational strength." As we will see, these dynamics have been shown to increase the likelihood of crime in such neighborhoods.

The three key issues that have been central to the preceding discussion, that is, neighborhood, a systemic approach to control, and crime, form the framework around which the remainder of this book is written. However, as is often the case in the social sciences, terms that seem to be self-explanatory are often very difficult to define precisely. Although readers of the opening section have all been exposed to the same words, there may be a great variation in the images of neighborhood, control, and crime that have been evoked. Therefore, it is necessary to specify clearly what it is we have in mind in our subsequent discussions.

The Neighborhood

What Is a Neighborhood?

Our favorite definition of the neighborhood comes from an old newspaper column written by Murray Kempton which appeared in the December 3, 1957, edition of the *New York Post* (see Kempton 1963). In that piece, Kempton expressed his frustration at the inability of a panel of experts called together by the Carnegie Institute to agree on a working definition of the neighborhood. However, Kempton was struck by the description offered to him in private by a worker for the Puerto Rican Labor Office: "A neighborhood is where, when you go out of it, you get beat up."

Although there still is not a consensus within sociology about how best to define this central aspect of American life, several key themes have emerged around which there appears to be general agreement (see the discussion of Hallman 1984:Chapter 1). First, and most basically, a neighborhood is a small physical area embedded within a larger area in which people inhabit dwellings. Thus, it is a geographic and social subset of a larger unit. Second, there is a collective life that emerges from the social networks that have arisen among the residents and the sets of institutional arrangements that overlap these networks. That is, the neighborhood is inhabited by people who perceive themselves to have a common interest in that area and to whom a common life is available. Finally, the neighborhood has some tradition of identity and continuity over time.

American sociologists have been fascinated with the origins and social life of local neighborhoods almost since the time that the discipline was first formally organized into a department at the University of Chicago in 1892. Contemporary urban sociology continues to be influenced strongly by the seminal work of two of the primary figures of the "Chicago School": Robert E. Park and Ernest W. Burgess.

The urban work of Park and Burgess was grounded in the assumption that competition was the fundamental form of social interaction that determined the territorial distribution of populations (1924:506–508). A key aspect of this competition was a struggle over the freedom to occupy and control scarce but desirable physical space, much in the same way that plants and animals compete for scarce resources. This similarity of the processes underlying their characterization of human patterns of locality and those found in the plant and animal kingdoms was intentional and led to the labeling of their approach as "human ecology" (see Hawley 1950).

Since desirable space is a scarce commodity, Park and Burgess assumed that it was subject to the same laws of supply and demand as other valued resources. Thus, they argued that a "biotic order" existed that reflected the dynamics of the competitive market system and, in turn, resulted in the existing pattern of land usage and the spatial location of population groups, In such a market, the price of housing reflects the relative demand for a particular property or area in the city; consumers in such a market may change residence as often as they wish if their income and credit worthiness makes such a move affordable (see Bottoms and Wiles, 1986; Bursik 1989).

This orientation is best reflected in Burgess's concentric zone theory of urban structure (1925). According to this argument, the most desirable (and, therefore expensive) land values were at that point where lines of transportation converged. This was usually in the center of the city where, owing to access to these lines, most of the commercial activities of a city were concentrated. In anticipation of the physical growth of this central business district, real estate speculators would purchase relatively inexpensive land directly surrounding the area in hopes of significant future profits. Since the maximization of profits entails the simultaneous minimization of costs, these speculators spent very little money for the upkeep of this property. As a result, the dwelling units in this area were usually in a state of disrepair and had relatively low rental and property values. Therefore, the areas immediately surrounding the central business district were the least attractive in the city and, owing to the presence of inexpensive housing, functioned as the typical initial area of residence for immigrant ethnic groups.

As these immigrant groups became more fully integrated into the economic structure, they were assumed to move progressively outward from the central city into more attractive and more expensive housing units. Areas that were least attractive (i.e., close to the central business district) tended to be characterized by high rates of population turnover, as residents moved out of them as soon as economically feasible. In addition, since this rapid transition made it difficult to form strong formal and informal linkages among the residents, it was very difficult to control the movement of unwanted new residents into the area. Therefore, these neighborhoods also were characterized by relatively high rates of population heterogeneity (see Bursik 1986a, 1989).

Park and Burgess also argued that a second set of dynamics, which they called the "moral order," was interdependent with the biotic dynamics of the competitive market system. This order reflected "problems of accommodation, or articulation, of groups within the community and of the adjustment of the life of the local community to the life of the wider community of which it is a part" (1924:720). The social contact between groups that initiates this accommodation was felt to create "sympathies, prejudices, [and] personal and moral relations which modify, complicate, and control competition" for desirable space (p. 507). Nevertheless, the argument of Park and Burgess gave the cultural and symbolic dynamics a role in the overall scheme of urban dynamics that is decidedly secondary to the biotic order. As a result, the moral dimension of urban life has not been a major component of most studies grounded in the theory of human ecology.[5]

Given these dynamics, neighborhoods were considered to be the result of the selective movement of the population into areas associated with particular economic, cultural, or occupational groups (Burgess 1925:54). Since these geographic groupings were assumed to be the result of the "natural mechanisms" of the competitive market system, they were generally referred to as "natural areas" (see Park 1926) and were considered to be "little worlds . . . each one differentiated from the others by its characteristic function in the total economy and cultural complex of city life" (Wirth and Furez 1938).

The notion of natural neighborhood areas became a central component of the empirical research being conducted in Chicago. Seventy-five natural areas were delineated by the Social Science Research Committee of the University of Chicago with the cooperation of many local agencies and the

Bureau of the Census. The chief considerations in the determination of the boundaries of these areas, which came to be known as the "official" local community areas of Chicago, were the settlement, growth, and history of the area, local identification with the area, the local trade area, the distribution of membership of local institutions, and the presence of natural and artificial barriers, such as rivers and highways (see Kitagawa and Taeuber 1963). *Local Community Fact Books* have been published roughly every ten years in Chicago, and the boundaries and names given to many of these areas almost seventy years ago still are the basis of identification and sentimental attachment for many Chicago residents. Attempts to officially delineate similar urban neighborhoods have been conducted in many other cities.

In addition to the assumption that such natural areas initially arise on the basis of competition within the housing market, a second characteristic of this approach is the assumption that neighborhoods have a dynamic character. Burgess (1925) argued that physical expansion was a central feature of modern urban life. Therefore, there is a tendency for the centralization of populations into a geographic area to be followed by a period of decentralization during which these groups attempt to move into adjacent neighborhoods. This has been referred to as the invasion/succession process and has typically been used to study the dynamics of change in the racial or ethnic composition of neighborhoods. Although there are several competing paradigms concerning the specifics of this process (generally referred to as the life-cycle, arbitrage, and composition models; see Schwab 1987), all share the image of the neighborhood as an area of ongoing change and adaptation.

The natural area approach continues to be very influential in the field of urban sociology, and many studies of neighborhood and crime to be discussed in this book use the original natural area boundaries of Chicago to determine the local communities under analysis. However, several serious criticisms have been made of this approach.

The first concerns the assumption that neighborhoods arise and develop solely on the basis of open, competitive market dynamics. Many writers have observed that the effects of "nonnatural" market mechanisms on the spatial location of population groups have been especially pronounced in the United States since World War II (see the discussion of Bursik 1988). Hirsch (1983), for example, has presented strong evidence that the activities of slumlords in Chicago's traditional Black Belt accelerated movement out of that area over what might have been expected given the economic status of the residents.

This manipulation of a neighborhood's housing market has not been solely determined by private initiative. As Guest notes (1984:293), the large bureaucracies that have arisen since World War II "undoubtedly have important influences over the political processes in determining the allocation of land." One of the reasons for the increase of such bureaucracies is that with the rise of suburbanization and the resulting decline in the population of many central cities, local governments are finding themselves facing extreme fiscal strain. A common response to this crisis has been the creation of zoning regulations that attempt to simultaneously maximize the tax yield from the properties in a neighborhood and minimize the public dollars necessary to service the community (Foley 1973:111).

In addition, incentives have been offered to potential builders and developers within some neighborhoods that were not necessary in the past. As Suttles (1972:82–86) has indicated, current decisions to develop an area are based not simply on an economical use of land, but also on expectations concerning the future potential of adjacent property. Since few developers or realtors are large enough to control such a large block of land and since many private firms are reluctant to risk a major investment in an area in which the future is problematic, the local government is forced to provide inducements to such developments, such as the financing of construction, the clearance and sale of land, and the establishment of standards for builders (p. 82).

Such incentives also have been used to influence the amount of residential mobility among neighborhoods. Clarke and Moore (1980:14–15) note that these may include the provision of grants to increase the perceived benefits of moving to or remaining in a particular neighborhood (such as for the rehabilitation of older dwelling units), the direct control of the relative costs of alternative dwellings (through rent control and the manipulation of interest rates), and attempts to influence the cost of moving (such as through subsidies and capital gains taxes). Bursik (1989), for example, presents evidence that the political decision to locate public housing within particular neighborhoods tends to increase the residential instability of those neighborhoods. If many former residents abandon existing local institutions, they must be created anew or face extinction. In addition, the linkages between residents that are essential for the development of community control are difficult to create and maintain during periods of rapid population turnover (see Bursik and Webb 1982). Therefore, although the traditional assumption of human ecology that a free and open competitive housing market gives rise to a set of "neighborhoods" is partly true, it is an extremely limited and incomplete perspective on a complex phenomenon.

The second, and related, criticism of the natural area approach concerns the relative inattention that traditional human ecology has paid to the degree to which the identification of local neighborhood is shaped by cultural and symbolic factors. As noted by Hunter and Suttles (1972:47), Park and Burgess assumed that the process of ecological sorting that shaped the neighborhood structure of a city resulted in the development of social bonds that drew everyone in an area together into a single community in which they shared a "common residential identity, sense of solidarity, and willingness to cooperate." Yet Hunter (1974) has shown that these ecological dynamics are in a continual interplay with symbolic issues that reflect changes in the spatial distribution of racial and economic populations and the tendency of residential groups to define themselves in terms of relative differences from other groups (see also Hunter and Suttles 1972:50). Thus, while there has been an overall persistence in the relevance of the 75 official community areas defined by Burgess in Chicago, ambiguity has also arisen concerning the names attached to particular neighborhoods and the boundaries that define those neighborhoods; Hunter (1974) presents evidence that 206 smaller, but meaningful, neighborhoods are embedded within these 75.

In addition, Hunter and Suttles (1972:45–61) also note that neighborhood residents live within a "pyramid of progressively more inclusive groupings," or what Jeffrey Slovak (1986) has referred to as "nested" communities. The smallest of these represents a network of acquaintanceship based simply on the propinquity of residence and the common use of local facilities (such as grocery stores). They refer to this as the "face-block" level since the physical residential block is a key source of such networks. This neighborhood, in turn, is embedded within a "nominal community," which is the smallest entity with a name known and recognized by both its residents and outsiders; Hunter and Suttles argue that it represents a haven of safety and identification to those living there.[6]

The third level is the "community of limited liability," a label derived from the seminal work of Morris Janowitz (1951). Such communities have institutionalized boundaries and officially recognized names and identities. In some cities, such as Chicago, the nominal community and the community of limited liability may be identical. However, there are many communities of limited liability within an urban system, such as police and school districts, geographically oriented religious congregations, and political wards. Not only do urban residents therefore live in several such communities simultaneously, but the boundaries do not typically coincide with each other. A resident's identification with a particular community of limited liability is partial, limited, and dependent on the current salience of issues being raised within the boundaries of those communities. Therefore, such boundaries, while institutionalized, are especially fluid in terms of their relevance to a person's identification with that neighborhood.

The fourth level discussed by Hunter and Suttles is the "expanded community of limited liability," which represents fairly large geographic areas of the city, such as "the South Side" or "the North Side." The identification with such a large area as one's neighborhood is much more limited than any of the other three levels and might pertain to such issues as perceived differentials in the services provided to those areas. However, despite the partial base of involvement with a large part of town, the identification with that area can be very intense on even that limited basis. Hirsch (1983), for example, has documented the sometimes intense rivalry between the black populations residing on the West and South Sides of Chicago. Even in cases that may appear fairly trivial, such as when different athletic teams are identified with particular parts of town (such as was the situation with the Yankees, Dodgers, and Giants in New York and is currently the case with the White Sox and Cubs in Chicago), these large geographic rivalries can be the source of violence.

Such considerations make it clear why it is very difficult to arrive at a simple definition of the neighborhood. Most of the theoretical arguments that will be developed in the subsequent chapters of this book focus on dynamics that are assumed to exist at the face-block and nominal community levels described by Hunter and Suttles. However, as we will see, there is a great deal of variation in the way in which "neighborhood" has been operationalized in the research that we will discuss. At one extreme are those survey-based studies that simply ask questions concerning the state of affairs in the respondent's neighborhood and leave it to the respondent to define what that means. At the other extreme are those studies that utilize data that have been collected by the Bureau of the Census or local public and private agencies. In this situation, there is much less flexibility in the delineation of the neighborhood, leading to operational definitions based on face-blocks, census tracts, or some aggregation of those tracts. In some cities, there is evidence that these units of analysis have some face validity as local community areas. In others, this must simply be assumed.

An awareness of such differences in the definitions of the neighborhood that are used in the research to be discussed is critical to a full understanding of the implications of that research due to the statistical problem known as "aggregation bias" (see Moorman 1980). Simply stated, this means that the findings that result from a statistical analysis are partly dependent on the size of the unit of analysis that is used. This was illustrated strikingly in an important paper presented by William Bailey at the 1985 meetings of the American Society of Criminology. Bailey took a single dataset, recoded it to represent the crime rates in U.S. states, Standard Metropolitan Statistical Areas (SMSAs), and cities, and examined the social, demographic, and economic correlates of those crime rates. Very different patterns emerged among the three different analyses, including changes in the direction of the relationships. He concluded that the "most appropriate" unit of analysis could only be determined on the basis of the theory underlying the research.

With very few exceptions, almost all the research to be discussed in this book was conducted on the basis of indirect approximations of the orientation to the neighborhood that we have developed in this section. Therefore, it is possible that in some cases, differences that have been found among studies may simply represent the effects of the different definitions of the neighborhood that have been utilized. As a result, the reader should take care to keep those definitions in mind. On the other hand, this limitation has a surprising strength. In many cases, we will see that studies based on very different operational definitions of the neighborhood have come to strikingly similar conclusions concerning certain aspects of the relationship between neighborhoods and crime. Such congruence suggests that we can place a great deal of confidence in those findings and may generalize from them in the development of our discussion.

The Systemic Theory of Neighborhood Organization

As noted previously, the primary orientation of this book is on the ability of neighborhoods to control themselves and their environment through formal and informal relational networks so that the risk of crime is minimized. Therefore, we need to understand just how such control might be possible in a modern, urban setting.

Some early sociological approaches to urban life presented a very pessimistic view of the ability of neighborhoods to act as an agency of social control. The most famous example of this orientation can be found in Louis Wirth's essay "Urbanism as a Way of Life" (1938), in which he argued that the growth in population, density, and heterogeneity that characterized urban areas led to a "segmentalization" of life, that is, a tendency to develop secondary rather than primary relationships (p. 12). As a result, interactions among city dwellers tended to be impersonal, superficial, transitory, and exploitative, for, as he argued, "the role which each one plays in our life is overwhelmingly regarded as a means for the achievement of our own ends."

Wirth's bleak image of urban life portrays the city dweller as relatively unconstrained by the personal and emotional controls of intimate groups. As a result, this lack of primary group affiliations, combined with a high rate of residential mobility, meant that "only rarely is he a neighbor" (p. 17). This portrayal is in stark contrast to the argument of Park and Burgess that the ecological dynamics of a city would generate a sense of collectivity and solidarity within urban neighborhoods. However, despite the strong influence of the Park and Burgess human ecology model on the development of urban sociology in the United States, their arguments concerning bonds of attachment among the residents of urban neighborhoods were largely ignored. As a result, the Wirth characterization of urban life dominated the literature for many years and as recently as 1977 was called "the single most widely accepted theory of the effects of urbanization on human behavior" (Berry and Kasarda 1977:55).

Nevertheless, especially since the publication of Morris Janowitz's *Community Press in an Urban Setting* in 1951, a large body of research has appeared that strongly challenges Wirth's image of the anomic urban dweller. Rather, this "systemic" approach to urban organization considers the local community to be a "complex system of friendship and kinship networks, and formal and informal associational ties rooted in family life and personal socialization requirements" (Berry and Kasarda 1977:56).

In our opinion, the systemic approach became a major alternative to the Wirth perspective with the publication of a study by John Kasarda and Janowitz in 1974. Contrary to Wirth, they argued that population size and density would not be associated with differences in participation and attachment to one's local community. Rather, the length of time that a person had resided in an area was the key consideration. Since it takes some period of time to develop extensive friendship, kinship, and associational ties within a neighborhood, people would become increasingly embedded within the local networks of affiliation over time. In turn, these attachments would foster increased levels of identification with and positive sentiment directed toward the neighborhood. Finally, they proposed that these proceses might also be shaped by the social status of the resident and his or her stage in the life cycle.

Although they presented mixed findings concerning the role of social status and age, their results overwhelmingly supported the argument concerning the role of length of residence: it was positively related to the number of friends, relatives, and acquaintances who lived in the neighborhood, participation in local organizations, and sentiment toward the community. On the other hand, the size and density of the area had a minimal effect on the process. Recent work, such as that of Lewis and Salem (1986), Sampson (1988), and Tittle (1989), has provided strong confirmation of the role of length of residence on the extent of a person's involvement in the social life of the neighborhood.

As will be seen in the subsequent chapters, the systemic theory is a central part of the framework that we will use to understand the relationship between neighborhoods and crime and is intimately connected with a perspective on neighborhood crime that has a long tradition in the field of sociology: the theory of social disorganization (see Chapter 2). The linchpin of this connection concerns the ability of a neighborhood to control the level of crime within its boundaries.

Systemic Control

Defining Social Control

For those of us who attended college and graduate school during the turbulent years of the 1960s and 1970s, there was a great deal of ambivalence and, for some, resistance to the key sociological notion of social control, for it seemed to imply the rigid suppression of nonconformity and the imposition of a set of dominant norms, values, and beliefs with which many of us were uncomfortable. Although this was an unfortunate caricature of what Gibbs (1989) has referred to as the central issue of sociology, it was not totally unwarranted, for such themes did, in fact, appear to be implicit to some of the classical statements of the problem, such as those presented by Ross (1901) and Sumner (1906). As Coser (1982) has observed, much of this writing assumed that humans were born with an innate nature or personality that was in conflict with the overriding needs of the social group. Therefore, for the group to survive, legal sanctions were needed to control these "drives" (also see Black 1989).

In the early decades of this century, sociologists began to downplay the role of innate natures and focus on socialization as the primary means of control. As a result, theories of social control had an overwhelmingly normative emphasis that is still the dominant concern of many current, albeit significantly more sophisticated, approaches. For example, Donald Black (1989) has recently discussed social control in terms of how people respond to deviant behavior, which he defines as "conduct regarded as undesirable from a normative standpoint" (p. 5, footnote 7).

While the enforcement of normative expectations certainly is an important dimension of social control, a strictly normative orientation imposes severe limitations on our understanding of that phenomenon. The complexities of the control of behavior perhaps are illustrated most strikingly in the unusual alliances that have developed in the effort to eliminate or severely restrict access to material that is considered by some to be pornographic. For example, temporary coalitions developed among certain feminist blocs and some extremely conservative political groups (such as the Moral Majority) to lobby for the passage of antipornography legislation in Minneapolis and Indianapolis. Although they had a mutual concern about the effects of pornography, it certainly cannot be argued that these groups shared a broad common base of values and beliefs, especially pertaining to the role of women in society. Likewise, some religious denominations with very different theological and political orientations (such as the Church of God and the United Methodist Church) have temporarily united around this issue (see Franc 1986). Therefore, the image of social control as the imposition of a monolithic belief structure on a nonconforming population is misleading.

A very different image can be found in the work of Park and Burgess (1924:Chapter VII) in which they describe social control as the regulation of behavior so that corporate action on behalf of the group is possible. The implications of this orientation have been most fully developed by Janowitz (1976, 1978), who defines social control as "the ability of a social group or collectivity to engage in self-regulation" (1976:9–10). Although much of the argument of Janowitz has a normative orientation, he does not consider social control to represent rigid repression. Rather, unconforming behavior can be tolerated as long as it does not interfere with the attainment of a commonly accepted goal. Therefore, there is no need for social control to imply a set of high moral and ethical principles to which groups should aspire (Janowitz, 1976:9), nor is it necessary to accept the assumption of many early criminologists that the members of a group subscribe to (or are expected to subscribe to) a set of "universal human needs" (Kornhauser 1978:63)

In this book we assume that the residents of neighborhoods share a common goal of living in an area relatively free from the threat of crime (see Bursik 1988). Therefore, social control represents the effort of the community to regulate itself and the behavior of residents and visitors to the neighborhood to achieve this specific goal. While this is a normative assumption, we feel that it is one supported by a great deal of evidence, as will be discussed in the "Crime" section of this chapter.

Thus, the arguments to be developed in subsequent chapters of this book are based on an assumption that the central underlying dynamic of neighborhood social control is the attempt to protect the area from threats that may undermine its regulatory ability. The most obvious dimension of crime-related threat is a physical one, directed not only at oneself, but also at one's family members. It is difficult to overstate the daily problems of safety encountered in some central-city neighborhoods of the United States. This was powerfully illustrated in a 1987 *Wall Street Journal* that which focused on Lafayette Walton, a 12-year-old boy living in the Henry Horner housing project in Chicago, whose daily existence was replete with gun battles, gang warfare, homicides, rapes, assaults, and exposure to serious drug abuse (Kotlowitz 1987). Although this is an extreme example, recent victimization data published by the Bureau of Justice Statistics (1990) indicate that children between the ages of 12 and 14 are victims of crimes of violence and theft at significantly higher rates than adults who are over the age of 25. Relatedly, Zinsmeister (1990) reports that almost 300,000 high school students are physically attacked each month. Thus, it is a common finding of residential mobility studies that families with children are more likely than others to move out of a neighborhood because of its high crime rate (see, for example, Katzman 1980).

Perhaps equally obvious are the symbolic threats that some criminal behaviors pose to the prevalent expectations concerning the rules of social life within the neighborhood. As Jack Douglas (1970:vii) has noted, "only shared rules, which are essentially prescriptions and proscriptions of typical actions in typical everyday situations . . . have proven capable thus far of producing the degree of ordering of interactions which human beings have found necessary for existence and for the good life." When such rules are violated with seeming impunity, a core aspect of the moral order underlying the neighborhood's collective life may be called into question.

However, criminal behavior may also threaten some less obvious aspects of neighborhood life. Sheldon Eckland-Olson (1989:209–210) has argued that processes of social control may be activated when any existing relational network is confronted with any possible disturbance. For example, the findings of Skogan (1990) suggest that residents of crime-ridden areas often withdraw from participation in local networks and activities owing to a fear of victimization and a distrust of fellow neighbors. Thus, the threat of crime may disrupt the effective ties that bind together the residents of an area.

Eckland-Olson also notes that threats to the existing networks of exchange may elicit social control. In the context of our argument, these exchanges would be primarily economic in that fewer people may feel comfortable entering the area to patronize local business; likewise, if the neighborhood is perceived as an undesirable residential area owing to its crime problems, there may be a decrease in the revenue that can be generated by the local housing market (see, for example, Schuerman and Kobrin 1986). In addition, an increase in crime or a change in its qualitative type may also threaten accommodations that some neighborhoods have made with certain criminal enterprises within the area from which it benefits economically. As Bernard Cohen (1980) found, in his study of prostitution in New York City, potential clientele are often hesitant to patronize services located in high-crime areas, owing to a fear for their own personal safety and fear of detection by the increased police surveillance assumed to exist in such areas. Thus, an increase in certain kinds of crime in the area would threaten the financial status of businesses located in these neighborhoods, such as hotels, coffee shops, and taverns (see Skogan, 1990:62). The forms of social control that may emerge when the economics of prostitution in an area are threatened by the presence of crime have been vividly described by Prus and Irini (1980).

The Processes of Social Control at the Neighborhood Level

Albert Hunter (1985) has developed a three-level approach to local community social control that provides a useful framework for understanding how relational networks are intrinsic to the control of crime at the neighborhood level. His most basic order of control is at the "private" level, which is grounded in the intimate informal primary groups that exist in the area. Within such groups, social control is usually achieved through the allocation or threatened withdrawal of sentiment, social support, and mutual esteem (p. 233). Black (1989:4) has provided a good summary of some of the primary mechanisms of control that may exist at this level, such as direct criticism, ridicule, ostracism from the group, deprivation, a resort to third parties, desertion, self-destruction, or violence.

Research that has been conducted within the deterrence framework of criminology has provided a large body of evidence concerning the effectiveness of such mechanisms in the control of crime (see, for example, Paternoster et al. 1983). However, two recent papers suggest that this form of social control might be more effective for adolescents than adults. Williams and Hawkins (1989), for example, find that the threat of social disapproval due to an arrest for wife assault did not have a deterrent effect on the behavior of adult males. Likewise, Grasmick and Bursik (1990) have presented evidence that the expected disutility due to losing respect from significant others does not have a deterrent effect on the inclination to cheat on one's taxes, steal, or drive under the influence of alcohol. There are important differences in the measures used in these studies that do not make the results directly comparable. Nevertheless, such conflicting results indicate that the dynamics of private social control are much more complex than often assumed.

The second level of control discussed by Hunter is called the "parochial" order and represents the effects of the broader local interpersonal networks and the interlocking of local institutions, such as stores, schools, churches, and voluntary organizations (p. 233). That is, whereas the private order of control refers to relationships among friends, the parochial order refers to relationships among neighbors who do not have the same sentimental attachment.

There is no question that this order has an important role in attempts to control crime in the neighborhood, as will be discussed in detail in our chapter on local community organizations. However, Spergel and Korbelik (1979:109) have shown that there are externally determined contingencies that mediate the ability of local networks and institutions to control the threat of crime. In fact, some local associations initially arise due to the intervention of external organizations who may seek legitimacy for projects they are considering in the area (Taub et al. 1977). Therefore, it is also necessary to consider the "public" level of social control (Hunter 1985:233), which focuses on the ability of the community to secure public goods and services that are allocated by agencies located outside the neighborhood.

These external resources can take two basic forms. First, the neighborhood may wish to engage in crime control activities that can only be successful if local organizations and representatives have the ability to influence municipal service bureaucracies and public/private decision-making agencies to allocate economic resources to indigenous organizations (Lewis and Salem 1986:79). For example, a neighborhood may attempt to raise funds for the creation and maintenance of a local social service agency; the success of many such programs is absolutely conditional on the ability to solicit external financing.

The second, and perhaps most important, external resource concerning the control of crime concerns the relationships that exist between the neighborhood and the police department of the city in which it is located. Not only are the residents of a neighborhood affected by police activities occurring within its boundaries, but studies of criminal decision making have shown that potential offenders often choose the area in which to commit a crime on the basis of the differential patterns of law enforcement in a city (see Carter 1974; Carter and Hill 1978; Rengert and Wasilchick 1985). As a result, when police activities are increased in one area of the city, there may be a tendency for crime rates to increase in adjacent neighborhoods where the risks of apprehension are not so great (see McIver 1981). The potential for such crime "spillover" suggests that the nature of police-community relations in a particular neighborhood is partly a function of simultaneous police activity in other nearby areas.

This issue will be extensively addressed in our chapter on local community organizations. In this introductory chapter, it may be sufficient to note that the problems of police estrangement from local communities have been widely recognized and discussed (see Skogan 1990). However, it is worth pointing out that inconsistent evidence exists concerning the influence of neighborhood characteristics on the delivery of police services (compare Smith 1986 with Slovak 1987).

The relationship of a systemic approach to neighborhood organization with the three-tiered approach to social control that characterizes this book is straightforward. As noted in our section on the neighborhood, the private, parochial, and public networks capable of social control do not develop instantaneously. Rather, they slowly emerge through interaction among the residents over a period of time. Therefore, the greater the level of residential instability that exists in a neighborhood, the less likely it is that such networks are able to control the threat of crime in an area since, as Bursik and Webb (1982:39–40) argue, ongoing instability makes it difficult to establish informal and formal associations that can be maintained over time.

Unfortunately, as we will see, the same kinds of definitional problems that characterize the operationalization of neighborhoods also plague much of the research pertaining to internal control. In general, studies that are based on survey research designs have been much more successful in obtaining data pertaining to relational networks that exist in the neighborhood than those based on data obtained from law enforcement and other agencies. As a result, some studies have very weak and indirect measures of social control, while others are only able to analyze processes that are logically related to (but conceptually distinct from) such control. Although there are some important exceptions, we think that it is fair to state that of the three central themes of this book, the empirical findings concerning social control at the neighborhood level easily are the least sophisticated.

Crime

What Is a Crime?

Unlike some other substantive areas examined within the social sciences, one of the most problematic issues in the area of criminology actually involves defining the behavior that we study. While this may seem to be a curious state of affairs to an outside observer, some very tricky conceptual and theoretical problems arise in this seemingly simple task. The traditional approach has assumed that a broad degree of consensus exists concerning the distinction between criminal and conventional behavior; this is usually referred to as the normative orientation. From this perspective, criminal law is simply a formalization of an underlying moral order subscribed to by most people.

Beginning with the pioneering work of Sellin and Wolfgang (1964) on the perceived seriousness of crime, a great deal of support has been provided for the normative orientation. For example, the National Survey of Crime Severity (Wolfgang et al. 1985) indicated that there was wide agreement that the most serious crimes reflected the general categories of homicide, rape, robbery, assault, and kidnapping, followed by crimes against property. Reid (1988:25) refers to such crimes as "mala in se," that is, perceived as intrinsically evil.

However, some strong objections have been made of the consensual assumptions that characterize normative approaches to crime. For example, the definitions of some behaviors generally considered to be criminal actually show a great deal of variation over time and space. An example that is often used in introductory crime, delinquency, and deviance courses presents the students with a scenario in which three women with purses are walking down a street. One of the purses contains an ounce of gold, one contains an ounce of marijuana, and one contains a pint of bourbon. The class is then asked how many of the women are breaking the law. The answer, of course, is that it depends, for over the course of the history of the United States, possession of each of the substances has been both legal and illegal.

The purpose of such an illustration is that normative orientations take the definitions of crime for granted, thereby ignoring the social dynamics and negotiations that gave rise to the illegalization of the behaviors in the first place. As a result, an alternative approach to the definition of crime, sometimes referred to as the relativistic orientation, has become increasingly popular, especially since the 1960s. One of the hallmarks of this orientation is an emphasis on the diversity that may exist in an advanced capitalist society such as the United States. From this perspective, criminal law is viewed as the outcome of a struggle between competing groups to formally protect the legal and institutional primacy of their normative standards, the symbolic status of their life-styles, or their present and anticipated economic resources.[7]

One of the best illustrations of these processes has been provided in Patricia Morgan's (1978) discussion of the passage of the 1881 California law that made opium smoking a misdemeanor. Although the use of opium and its derivatives was relatively common in the second half of the nineteenth century, smoking the substance was associated primarily with the Chinese working class who had emigrated to the United States. Given the poor economic situation in the United States following the Civil War (culminating in the depression of 1871), this population posed a major threat to the already dwindling economic resources of American labor owing to its willingness to work for fairly low wages. Initially, efforts to control this threat took the form of immigration restrictions and exclusionary laws directed against the Chinese. However, the conflict rapidly developed symbolic dimensions in which the Chinese were portrayed as posing a threat to the American life-style. For example, Morgan (p. 57) cites an 1878 report of the San Francisco Police Department to the California State Senate Committee on Chinese Immigration in which it states that the department had found "white women and Chinamen side by side under the effects of this drug—a humiliating sight to anyone who has anything left of manhood."

As a result of such efforts, California passed legislation banning the smoking of opium. It is important to emphasize that the smoking of opium took place almost exclusively in Chinese-operated opium parlors; other forms of opium use that were more prevalent in the non-Chinese population were not affected by this legislation. Thus, this first statewide law banning certain forms of opium use was clearly an effort to control the threats posed by the Chinese working class.

Therefore, the relativistic approach to crime questions the underlying consensus and moral order that is assumed by normative approaches to be reflected in the law. For example, according to the findings of the National Survey of Crime Severity discussed earlier in this section, there is a significant amount of variation in the perceived seriousness of behaviors that are often referred to as victimless crimes. Reid (1988:25) has referred to such offenses as "mala prohibita," that is, perceived as evil because they have been legislated as such.

Obviously, the dynamics that lead to the designation of particular behaviors as crime represent a mixture of the normative and relativistic dimensions. This presents a problem for a systemic control approach to neighborhood crime, for a central assumption of such an orientation is that consensus exists among the residents concerning the goal of living in an area relatively free from the threat of crime. If a widespread consensus cannot be shown to exist, such as in the case of relatively minor drug use, then the framework is simply not viable.

The key to the resolution of this issue lies in the recognition that at the heart of both the normative and relativistic approaches, as well as in the systemic theory of control discussed above, is the notion of threat. Normative approaches assume that laws provide some degree of protection against threats to a single shared moral order, whereas relativistic approaches assume that complex societies are characterized by a variety of such orders, competing economic interests, and battles over the institutional means of socialization, all of which may be reflected in the criminal law as an effort to protect social, symbolic, and economic resources that may be threatened.

While we accept the relativistic arguments concerning the implications of diversity and crime, the widely replicated seriousness literature indicates that certain behaviors, that is, those that have been referred to as mala in se crimes, are perceived as threatening by virtually all competing interest groups. Therefore, our attention in this monograph is restricted primarily to those crimes for which widespread consensus has been demonstrated. These tend to roughly correspond to behaviors generally referred to as "index" (or Part I) crimes in the Uniform Crime Reports.

Two points should be made regarding this restriction. First, the systemic control model may not provide an especially powerful approach to certain extremely serious crimes. Schrager and Short (1980), for example, present strong evidence that some white-collar or "organizational" crimes (such as manufacturing and selling drugs known to be harmful, overcharging for credit, and price fixing) are considered by most of the public to be as harmful as many of the "street" crimes that are the focus of this book. There are reasons to expect that the systemic approach would not successfully predict efforts to control such crimes. Suttles (1972:59) has argued that actions on behalf of a community of limited liability tend to be specialized and oriented toward limited issues. Street crime committed within the boundaries of an area presents an immediate threat to the members of that community and provides a focus on which the residents can unite. At times, as for example in the case of a price-gouging supermarket or landlord, white-collar crime might also generate enough perceived common threat to set the processes of control into action. On the other hand, white-collar crime committed by residents of the community but having no widespread impact on that community may not be subject to the same regulatory processes.[8]

Second, there is a growing body of literature that broadens the focus of systemic control to encompass problematic behaviors that in themselves are not especially serious or even illegal, such as public drinking, vandalism, panhandling, and loitering groups of youths (see Wilson and Kelling 1982; Lewis and Salem 1986; Skogan 1990). Evidence is accumulating that such activities may serve as a sign to neighborhood residents that their community is getting "out of control" and may be headed for serious crime-related problems. The examination of reactions to such "incivil" (Wilson and Kelling, Lewis and Salem) or "disorderly" (Skogan) behavior has provided some important insights into the processes underlying neighborhood control, which will be discussed in Chapters 2, 4, and 6. Unfortunately, as noted by Skogan (1990:3–9), the identification of those activities that violate norms pertaining to the use of public space is especially subsceptible to a middle-class bias. As we will see, there is evidence that this bias may not be as problematic as is generally assumed. However, the reader must realize that the theoretical justification for a systemic control approach to these behaviors is much weaker than for the types of crime we have discussed.

The Measurement of Crime

In general, five basic approaches have been used to study crime at the neighborhood level: official statistics collected by law enforcement agencies that can be aggregated to the neighborhood level, self-reported studies of offending, self-reported studies of victimization and fear, ethnographic community studies, and most recently, records of citizen calls to the police for crime-related service. The theoretical and measurement problems associated with each of these forms of data are well known and will not be discussed here; the interested reader is referred to O'Brien (1985), Kempf (1990), MacKenzie et al. (1990), or Sherman et al. (1989) for full reviews of these issues.

However, it is important to note at the outset that three important differences exist in the way that crime has been operationalized in the research we will discuss. First, some studies have focused on the crime committed by residents of the neighborhood, and others have examined the neighborhood distribution of crimes regardless of the place of residence of the offender. While such information is complementary, each reflects the outcome of a different set of underlying processes.

Second, studies based on official statistics have not collected their data from consistent points in the law enforcement bureaucracy. The most commonly used approaches have utilized calls for service, crimes reported to the police, crimes for which arrests have been made, or the referral of a case to criminal or juvenile court. Given the primarily reactive nature of policing (Black 1970), it is obvious that neighborhood crime rates overwhelmingly represent those offenses that have been reported by citizens to their local departments. Thus, to the extent that nonreporting is prevalent within a jurisdiction, many offenses will remain "hidden" from official records and not be reflected in the rates used to estimate crime trends.

In addition, howver, many citizen complaints also are not reflected in these rates through a failure of the police officers to file a formal written report on the complaint. Black (1970), for example, notes that police officers officially recorded only 64 percent of the complaints observed in his study of Boston, Chicago, and Washington, D.C., in 1966. This ratio is similar to that observed by Sherman et al. (1989) in Minneapolis during 1986, where only 66 percent of all robbery-related complaints resulted in an official recording. Such discretionary practices make it difficult to derive reliable estimates of even the number of citizen complaints, much less the actual volume of crime that occurs in a locality during any particular time period.

The decision-making processes that occur at each stage of the law enforcement hierarchy progressively reduce the number of criminal events that are available for analysis in datasets collected at those stages. In addition, the characteristics of criminal events still active at later stages of the process may significantly differ from those at earlier stages owing to this filtering process. These inconcistencies often make the direct comparison of the results from two or more studies very difficult. Such problems will be noted as they arise.

Finally, although the title of this book suggests a focus on crime, some of the most important research that we will discuss is actually based on the neighborhood distribution of delinquency, that is, criminal activities engaged in by juveniles. Since the bulk of criminal activity is in fact concentrated within this age group, studies of crime in general and studies of delinquency in particular are not as incomparable as they may appear. Nevertheless, this distinction represents an important difference in the study designs of the research that will be discussed.

2

The Criminal Behavior of Neighborhood Residents

> The paradox of 32nd Street culture is that although these youth are the most serious violators of the most important legal norms of the larger society, they do not conduct their activities in a spirit of opposition to the culture of mainstream society
> —(Schwartz 1987:215).

One certainly does not need an advanced degree to conclude that all neighborhoods do not have the same rates of criminal behavior. However, two very different social processes may account for this phenomenon. As we described in the opening chapter, this book assumes that such differences can best be understood in terms of variations in the abilities of local communities to regulate and control the behavior of their residents. Thus, crime rates are presumed to represent an outcome of group dynamics operating at the neighborhood level.

On the other hand, as Sampson (1987a, 1989) has observed, neighborhood crime rate differences might simply represent the spatial distribution of individuals with particular social and demographic characteristics. If some of these characteristics are associated with a higher likelihood of crime, and if individuals with these characteristics are more likely to live in particular areas of the city, then it may not be necessary to consider the role of neighborhood dynamics at all. Rather, differences in neighborhood crime rates may simply reflect differences in the composition of the neighborhoods; areas with high rates of crime are those inhabited by relatively high proportions of crime-prone individuals. Such an orientation would require a significant shift in emphasis from the "kinds of neighborhoods" that have a high likelihood of crime, to a focus on the "kinds of people" who commit crime. Therefore, before examining the neighborhood dynamics that may inhibit the level of criminal behavior among area residents, it is necessary to determine whether such a discussion is even warranted.

24

Neighborhood and Compositional Effects

It is a well-known fact that delinquent behavior has a strong relationship with age; its prevalence tends to increase around age twelve, peak at ages sixteen or seventeen, and then declines (see Tracy et al. 1990). Let us assume that the average age of children in the Minoso Park neighborhood is thirteen, while in Gibsonville it is sixteen. While the delinquency rate should be much higher in the second community than the first, this difference is due strictly to the demographic composition of two areas and not to any processes operating at the neighborhood level. As this example illustrates, it may be the case that differences in local community crime rates primarily reflect the variation in the distribution of people who have individual characteristics associated with a high likelihood of criminal behavior (Wilson and Herrnstein 1985:291).

This issue has been especially problematic when examining the effects of the racial and ethnic composition of a neighborhood on the crime rate. It has been documented widely that blacks have offending rates several times higher than that of whites (see Sampson 1987a:95). However, there is a great deal of controversy about how such findings should be interpreted. At one end of the spectrum, there are those who would argue that blacks have genetic structures that may predispose them to crime (see the discussion of Karmen 1980). A high correlation between the percentage of a neighborhood that is black and the crime rate would therefore reflect the distribution of such individual dispositions in the composition of the neighborhood. At the other end are those who argue that such a correlation reflects the fact that many blacks are forced to live in community contexts that are conducive to crime. For example, neighborhoods with relatively large concentrations of black residents also tend to be characterized by high rates of unemployment (the correlation for Chicago during 1980 was 0.81). If these contexts were taken into account, it is assumed that the association between percent black and the delinquency rate would disappear.

The classic example of disagreement over this issue arose in response to work conducted by Clifford Shaw, Henry McKay, and their associates (1929, 1931, 1942). On the basis of data that spanned a thirty-year period, Shaw and McKay argued that the neighborhoods of Chicago were characterized by relatively stable delinquency rates over time despite marked changes in their racial and ethnic composition. Therefore, they concluded that the "higher rates of delinquents found among the children of Negroes, the foreign born, and more recent immigrants are closely related to existing differences in their respective patterns of geographic distribution in the city" (1942:162). That is, some neighborhoods were characterized by ongoing traditions of delinquent behavior regardless of their racial and ethnic composition. Their position on this issue is reflected most clearly in their statement that "diverse racial, nativity, and national groups possess relatively similar rates of delinquents in similar social areas. . . . [I]n the face of these facts it is difficult to sustain the contention that, by themselves, the factors of race, nativity, and nationality are vitally related to the problem of juvenile delinquency" (p. 162).[1]

Christen Jonassen took great exception to these conclusions in a paper published in 1949, citing problems in the statistical methods and research design of Shaw and McKay and arguing (p. 614) that it was "rather improbable" that nationality was not "vitally" related to juvenile delinquency. In their response to Jonassen, Shaw and McKay (1949) acknowledged the limitations inherent to their data. Nevertheless, they insisted that the reason that black youths had higher rates than white youths was due to the community context, noting that "it cannot be said that they are higher than rates for white boys in comparable areas, since it is impossible to reproduce in white communities the circumstances under which Negro children live. Even if it were possible to parallel the low economic status and the inadequacy of institutions in the white community, it would not be possible to reproduce the effects of segregation and the barriers to upward mobility" (p. 617).

Reanalyses of the Shaw and McKay data by Bursik and Webb (1982) produced a strong confirmation of their findings: changes in the racial/ethnic composition of Chicago's neighborhoods were not significantly related to concurrent changes in the delinquency rates of those areas prior to 1940. However, this was not the case in subsequent decades. Chilton (1987), for example, has presented evidence that 24 percent of the increase in homicide arrests and 45 percent of the increase in robbery arrests that characterized Chicago between 1962 and 1980 could be attributed to the increase of nonwhite men in the population.[2] He suggests (p. 197) that this might be due to the declining economic circumstances of the urban black population, noting that if urban blacks were similar economically to the suburban white population, race might be shown to be unrelated to crime. This illustrates the point made by Shaw and McKay, that is, that the racial composition of a neighborhood is so highly correlated with other social and economic characteristics of local communities that it is nearly impossible to reliably separate the effects of racial composition on crime from the effects that can be attributed to other aspects of the neighborhood.

Because of this confounding, Sampson (1989:5) laments that "it takes little time to think of ten competing reasons—both compositional and macrolevel—why factors like poverty rates and percent black might be correlated with aggregate crime rates." We will present one such explanation for the correlation that has emerged between the percent black (or nonwhite) in a neighborhood and that area's crime rate that is grounded in our systemic theory of neighborhood control. Nevertheless, the reader contemplating the research discussed in this (and following) chapters should draw conclusions concerning the role of the local community with caution since several methodological problems have led some to conclude that the magnitude of the neighborhood effect is "hard to estimate" (Wilson and Herrnstein 1985:311).

Many of these problems are due to the measurement issues discussed in the first chapter. There is a class of neighborhood properties that, by definition, do not represent a simple aggregation of individual attributes; these are commonly referred to as "emergent properties" (Harré 1981). The concept of emergent properties assumes that there are characteristics of the whole that are not manifested by any of its parts when considered independently (Harré:142) and can be traced to Durkheim's arguments concerning the existence of "social facts." One of the best illustrations of such properties in the field of criminology can be drawn from the classic study of Cloward and Ohlin (1960), which argues that the nature of the adolescent delinquency that occurs within a particular neighborhood will depend on the articulation between the structures of legitimate and illegitimate opportunities that exist in that neighborhood.

The relational networks associated with the control of crime within a neighborhood, the viability of local neighborhood organizations as agencies of formal and informal social control, the linkages between these organizations, the political power base of the neighborhood, and the relationship of the local community to the wider urban context are all prime examples of emergent properties. Unfortunately, it is difficult to collect reliable data that facilitate the development of indicators of these properties. As a result, most studies have been forced to rely solely on compositional variables that reflect the distribution of some individual characteristic among the residents of the neighborhood.[3]

For example, Sampson and Groves (1989) derive their measures of the systemic structure of control in a neighborhood by averaging the responses of residents to questions pertaining to the number of friends living in the neighborhood and attendance at meetings of local organizations. While we will see later in this chapter that their work represents one of the best large-scale approaches to neighborhood control found in the literature, it is based nonetheless on compositional measures in which the presumed systemic effect is a strict function of individual patterns of behavior. Therefore, it is not an overstatement to note that the predictions of the systemic model of crime have never been thoroughly or adequately tested owing to a lack of data on the emergent properties of urban neighborhoods.

Separating Neighborhood and Individual Effects

Most of the studies discussed in this book assume that compositional variables provide important (albeit incomplete) information concerning the normative or structural "climate" that is dominant in a neighborhood (see

Blalock 1984:359). However, one can only fully be confident in such an assumption if the statistical models used to examine the neighborhood/crime relationship simultaneously control for the compositional effect at the individual level of analysis; this approach is known as contextual analysis (see Boyd and Iverson 1979).

Although the statistical aspects of contextual analyses are typically very straightforward, the research design necessitates the collection of data from both a representative sample of neighborhoods and a representative sample of residents from within each of those neighborhoods. Given the high costs of such designs, there is a relatively small body of contextual research in criminology through which we may assess the relative importance of neighborhood and individual effects. However, the research that has been conducted suggests that certain crime-related dynamics operate at the neighborhood level that are not simple functions of the individual charcteristics of the residents of that neighborhood.

In 1962, John Clark and Eugene Wenninger published a research note that examined the relative effects of individual and neighborhood socioeconomic status on the delinquency rate found in four northern Illinois communities. Although they concluded that the social class of the individual was generally unrelated to delinquent behavior, they argued that "there are community-wide norms which are related to illegal behavior and to which juveniles adhere regardless of their social class origins" (p. 833). However, it is difficult to reach any firm conclusions concerning these community effects on the basis of this study since some socioeconomic categories of the respondents had to be combined or ignored because of the small sample sizes in three of the communities.

The more recent study of Ora Simcha-Fagan and Joseph Schwartz (1986) suggests that the relationship between neighborhood dynamics and the generation of delinquent behavior among adolescent residents is not as straightforward as has been traditionally assumed. Simcha-Fagan and Schwartz utilize census materials to create a series of compositional variables pertaining to the economic level and residential stability of 12 New York City neighborhoods in which they had collected survey data concerning the delinquency of 553 youths. In addition, they also interviewed the respondents' mothers to obtain information on the nature of the systemic relationships that existed in the area. Overall, they found that while the characteristics of these neighborhoods were related significantly to delinquency, the magnitude of the effect was relatively small after controls were made for the compositional characteristics of their respondents. We will examine this study in some detail later in this chapter.

The most conceptually and analytically sophisticated examination of the relative effects of neighborhood and individual dynamics is found in Stephen Gottfredson and Ralph Taylor's (1986) study of the offense histories of approximately 500 offenders who had been released from incarceration into one of 67 Baltimore neighborhoods. The postrelease behavior of these individuals was monitored for the occurrence of an arrest, the time elapsed until that arrest, and the nature of the illegal activity. To capture the overall effects of the neighborhood dynamics, they created a series of "dummy" variables to represent each local community.[4] After controlling for the individual characteristics of the sample, these neighborhood effects were significantly related to all indicators of recidivism.

Overall, the studies that have been conducted suggest that the neighborhood does have a significant effect on the probability of criminal behavior that is independent of the effects that can be attributed to the personal attributes of the residents of the community. While we assume that a focus on the neighborhood context is theoretically warranted and empirically justified by such findings, the reader should be aware that the magnitude of this effect relative to that which can be attributed to individual-level processes is not entirely clear and cannot yet be resolved on the basis of existing research. As noted above, one of the greatest impediments to this resolution has been the enormous costs involved in the design and implementation of studies that are capable of simultaneously collecting significant amounts of individual and neighborhood-level data. However, the Office of Juvenile Justice and Delinquency Prevention has coordinated the collection of exactly these types of data in a set of very ambitious studies being conducted in Denver (Huizinga et al. 1991), Pittsburgh (Loeber et al. 1991), and Rochester (Thornberry et al. 1991). Therefore, there is a strong possibility that this issue will receive an increasing amount of attention in the near future.

The Theory of Social Disorganization and Systemic Approaches to Crime

Our systemic orientation to neighborhoods and crime is firmly grounded in a long-standing criminological tradition called social disorganization, especially as developed by Clifford Shaw and Henry McKay (Shaw et al. 1929; Shaw and McKay 1942, 1969; McKay 1967). Such a statement is likely to elicit groans from those readers who are somewhat familiar with the disorganization perspective, yet who have failed to notice the developments in the area during the last decade or so. In fact, many criminologists view the concept of social disorganization as largely marginal to modern criminological thought. William Arnold and Terrance Brungardt (1983:113), for example, flatly dismiss its relevance to theories of crime causation since "clearly, as we usually think of it, it is not even a necessary condition of criminality, let alone a sufficient one." In fact, as recently as 1987, James Unnever's review of the James Byrne and Robert Sampson (1986) collection of papers on the social ecology of crime states that "Shaw and McKay's theory of social disorganization, which gave birth to this area of research, has been soundly dismissed" (p. 845).

As we shall see in this chapter, there is no question that certain aspects of the original Shaw and McKay thesis were very problematic. However, despite the appearance of occasional obituaries for the perspective, there has been a significant revitalization of the theory within (especially) the last ten years. Admittedly, some criminologists have expressed chagrin at this resurrection. For example, one of our favorite professional anecdotes pertains to a conversation in which we were engaged during an annual meeting of the American Society of Criminology during the mid-1980s, when we were informed that social disorganization was the "herpes of criminology . . . once you think it is gone for good, the symptoms flair up again."[5] Nevertheless, we feel that the reformulation of social disorganization within a systemic perspective provides fruitful and important insights concerning the relationship between contemporary neighborhood dynamics and crime.

Social Disorganization

The Original Shaw and McKay Thesis

In 1921, Clifford Shaw began collecting a set of data from the Cook County (Illinois) Juvenile Court, the Cook County Boys' Court, and the Cook County Jail that included the home address, the offense, the age, and the gender of the offender. Little progress was made in this endeavor until 1926, when the study became an integral part of the work conducted by the newly established Department of Research Sociology at the Institute for Juvenile Research in Chicago (see Shaw et al. 1929:ix). In 1927, Frederick Zorbaugh, Leonard Cottrell, Jr., and, most important, Henry McKay joined the staff of the Institute to work on this project.

In all, eight primary series of data were collected (see Shaw et al. 1929:Chapter 3):[6]

1. The male truants (aged eight to sixteen) brought before the Juvenile Court on truancy petitions between 1917 and 1927 ($N = 5,159$)

2. The alleged delinquent boys (aged ten to sixteen) dealt with by juvenile police officers during 1926 ($N = 9,243$)

3. A series identical to (2) for 1927 ($N = 8,591$)

4. The boys (aged ten to sixteen) referred to the Juvenile Court of Cook County between 1917 and 1933 ($N = 8,141$)

5. A series identical to (4) for 1900–1906 ($N = 8,056$)

6. The boys (aged seventeen to twenty) brought before the Boys Court of Chicago on felony charges between 1924 and 1926 ($N = 6,398$)

7. The adult offenders (aged seventeen to seventy-five) placed in the Cook County Jail during 1920 (N = 7,541)

8. The girls (aged ten to seventeen) brought before the Juvenile Court of Cook County (N = 2869)

The acquisition of these data was not considered to be a one-time effort of the Institute for Juvenile Research. Rather, the data collection was instituted as an ongoing part of the Institute's operation. For the purposes of this chapter, the most important additions were series identical to (4) for the years 1927–1933 (N = 8,411; Shaw and McKay 1942), 1934–1940 (N = 9,849), 1945–1951 (N = 8,041), 1954–1957 (N = 9,830), 1958–1961 (N = 14,167), and 1962–1965 (N = 12,844; all series subsequent to 1927–1933 discussed in Shaw and McKay 1969). Through these efforts, Shaw and McKay were able to compile a dataset containing the addresses of male adolescents referred to the Cook County Juvenile Court from 1900 through 1965; the magnitude of this effort has never even been approximated in any other criminological study of which we are aware.

The residential address of each individual reflected in this database was plotted (by hand!) on a base map of the city of Chicago (see Shaw et al. 1929:24) for a full description of the process) and then copied into outline maps of Chicago by means of a reflector and glass-top table (see Figure 2–1).[7] The rates of delinquency (defined in terms of the number of boys referred to juvenile court) were then computed on the basis of census tracts, the official local community areas of Chicago, and one-square-mile areas of the city, which was their most common operational definition of the neighborhood.

On the basis of visual inspections of these maps, supplemented by some rudimentary statistical tests, Shaw and McKay reached two very important conclusions in the 1942 monograph. The first has already been noted in the opening section of this chapter: the relative distribution of delinquency rates remained fairly stable among Chicago's neighborhoods between 1900 and 1933 despite dramatic changes in the ethnic and racial composition of these neighborhoods. The second conclusion laid the groundwork for their application of the social disorganization approach: delinquency rates were negatively correlated with distance from the central business district. Given the positive correlation of neighborhood economic status with distance from the center of the city, this finding indicated that delinquency rates were negatively correlated with the economic composition of local communities.

It is important to stress the central difference between their documentation of that negative association and their theoretical understanding of the relationship, for it has led to some very basic misunderstandings of the social disorganization perspective. Shaw et al. (1929) were greatly influenced by the human ecology perspective that was developing at the University of Chicago and argued that the dynamics of city growth discussed by Burgess

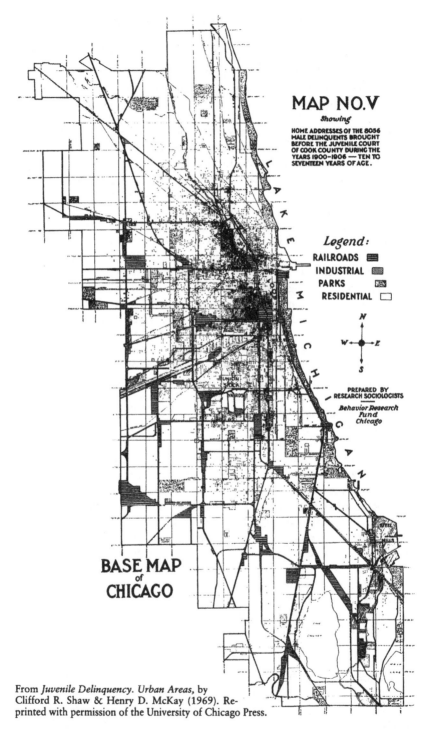

MAP NO.V

Showing

HOME ADDRESSES OF THE 8056
MALE DELINQUENTS BROUGHT
BEFORE THE JUVENILE COURT
OF COOK COUNTY DURING THE
YEARS 1900–1906 — TEN TO
SEVENTEEN YEARS OF AGE.

Legend:

RAILROADS
INDUSTRIAL
PARKS
RESIDENTIAL

PREPARED BY
RESEARCH SOCIOLOGISTS

*Behavior Research
Fund
Chicago*

BASE MAP
of
CHICAGO

From *Juvenile Delinquency. Urban Areas*, by
Clifford R. Shaw & Henry D. McKay (1969). Reprinted with permission of the University of Chicago Press.

Figure 2–1. Distribution of Male Juvenile Delinquents, Chicago, 1927–
33.

were "the frame of reference in which such a problem as juvenile delinquency may profitably be studied" (p. 21). This orientation was more fully developed in Shaw and McKay (1942) through a detailed discussion of the processes of invasion and succession, residential mobility, and the concentration of economic and occupational groups into particular neighborhoods in the city. Therefore, Shaw and McKay did not posit a direct relationship between economic status and rates of delinquency, although they are sometimes accused of doing so. Rather, they concluded that the pattern of neighborhood delinquency rates were related to the same ecological processes that gave rise to the socioeconomic structure of urban areas (see Bursik 1986a).

As we noted in Chapter 1, the ecological approach of Park and Burgess argued that areas characterized by economic deprivation tended to have high rates of population turnover because they were undesirable residential areas and people would abandon them when it became economically feasible. In addition, this rapid compositional change made it difficult for these communities to mount a concerted resistance against the influx of potentially threatening groups. Therefore, poor neighborhoods also tended to be characterized by racial and ethnic heterogeneity. Shaw and McKay argued that these characteristics made it very difficult for the neighborhood to achieve the common goals of its residents, a situation they called social disorganization, drawing from the seminal work of W. I. Thomas and Florian Znaniecki (1920).

The causal linkage between social disorganization and juvenile delinquency rates was not clearly explicated by Shaw and McKay. In various sections of their work, they freely draw on elements of strain, cultural conflict, and control theories (see Kornhauser 1978). However, given the intimate theoretical connection between processes of rapid ecological change and the social disorganization framework, a control-theoretic approach offers perhaps the best general basis for understanding the Shaw and McKay argument. Rapid population turnover and heterogeneity can decrease the ability of a neighborhood to control itself in order to provide an environment relatively free of crime for three basic reasons (see Bursik 1988:521):

1. Institutions pertaining to internal control are difficult to establish when many residents are "uninterested in communities they hope to leave at the first opportunity" (Kornhauser 1978:78).
2. The development of primary relationships that result in informal structures of neighborhood control is less likely when local networks are in a continual state of flux (Berry and Kasarda 1977).
3. Heterogeneity impedes communication and thus obstructs the quest to solve common problems and attain common goals (Kornhauser 1978:75).

Thus, weak structures of formal and informal control decrease the costs associated with deviation within the group, making high rates of crime and delinquency more likely. Framed in this manner, the social disorganization approach of Shaw and McKay is clearly a systemic theory of neighborhood crime control.

The Systemic Implications of the Basic Social Disorganization Model

One indication of the confusion that the concept of social disorganization has generated within criminology is the existence of at least five extended efforts to clarify the assumptions of the Shaw and McKay approach (Finestone, 1976; Gold 1987; Kobrin 1971; Kornhauser 1978; Short 1969). Of the four, Kobrin's is the most historically interesting because it originated as an internal memo circulated at the Institute for Juvenile Research while McKay was still employed there. This suggests that Shaw and McKay's model may not have been completely clear even to those people working with them.

The most important source of confusion concerning social disorganization is the fact that Shaw and McKay sometimes did not clearly differentiate the presumed outcome of social disorganization (i.e., increased rates of delinquency) from disorganization itself. This tendency led some to equate social disorganization with the phenomena it was intended to explain, an interpretation clearly not intended by Shaw, McKay, and their associates (1929:205–206). For example, Bernard Lander (1954:10) concluded that the value of the social disorganization construct is "dubious in view of the fact that social disorganization itself has to be defined as a complex of a group of factors in which juvenile delinquency, crime, broken homes . . . and other socio-pathological factors are included." Thus, Lander defined delinquency as social disorganization. It must be noted that Shaw and McKay were not totally responsible for this muddled distinction, for Stephen Pfohl (1985:167) has pointed out that the classic disorganization theorists of sociology often used a single indicator, such as a delinquency rate, as "both an example of disorganization and something caused by disorganization" (see his example concerning the mental health studies of Faris and Dunham 1939).

The modern reformulation of social disorganization as a systemic model of neighborhood control can clarify this distinction significantly. The instability and heterogeneity of local communities is assumed to affect the three levels of social control discussed by Hunter (1985). At the private level, ongoing changes in the residential population of a neighborhood make it very difficult to establish and maintain intimate primary ties within the community. Thus, affective relational networks tend to be fairly superficial and transitory, making the threatened withdrawl of sentiment, support, and esteem a relatively ineffective form of control. As Greenberg et al. (1985:46) have observed, while gossip, or the threat of it, is a powerful means of social control, its effectiveness depends on the inability of people to hide details of one's life from other members of the peer group. When the emotional relationships within this peer group are fairly superficial, this is easily accomplished.

While neighborhood instability makes the development of deep and lasting affective relational networks very difficult, heterogeneity in the area limits the breadth of such networks (see Gans 1962; Suttles 1968). This is nicely illustrated in Sally Merry's (1981) description of a public housing project populated by Chinese, blacks, whites, and Hispanics. Although over half the families had lived in the project for more than ten years and there was substantial daily contact among members of these groups when using the project facilities, friendship networks rarely crossed racial and ethnic boundaries. In such a situation, the overall capability of affective networks to control the behavior of the residents is extremely restricted. For example, Merry notes (p. 96) that when some of the black youths living in the project heard that the Chinese residents considered them to be criminals, "they were intrigued but unconcerned and made no effort to alter their behavior."

Recall that the second level of community control, which Hunter (1985) calls parochial, refers to relationships among the residents that do not have the same sentimental basis as the affective networks. Therefore, at this level of self-regulation, the systemic control of crime partly reflects the ability of local neighborhoods to supervise the behavior of their residents. Stephanie Greenberg and her colleagues (1982a, 1982b, 1985, 1986) have identified three primary forms of such supervision:

1. Informal surveillance: the casual but active observation of neighborhood streets that is engaged in by individuals during daily activities.
2. Movement-governing rules: the avoidance of areas in or near the neighborhood or in the city as a whole that are viewed as unsafe.
3. Direct intervention: questioning strangers and residents of the neighborhood about suspicious activities. It may also include chastening adults and admonishing children for behavior defined as unacceptable (1982b:147–148).

Instability and heterogeneity also weaken the supervisory capabilities of these parochial networks. Greenberg et al. (1982a) show that residents are not likely to intercede in criminal events that involve strangers and are reluctant to assume responsibility for the welfare of property that belongs to people they barely know. Therefore, supervision is less likely in areas in which there is not a relatively stable population base.

The social boundaries that may exist between groups in heterogeneous neighborhoods can also decrease the breadth of supervisory activities due to the mutual distrust among groups in such areas. Merry (1981:123) notes that robberies were committed within the housing project described above without fear of apprehension owing to the anonymity caused by the composition of the project. Mark Granovetter's (1973) discussion of urban networks suggests that in neighborhoods in which individuals are relationally connected to every other member of their network but to no one outside that network, supervisory activities have to develop independently within each network to ensure success in the control of crime throughout the community (see pp. 1373–1374). Thus, racial and ethnic heterogeneity may lead to a differential capacity of neighborhoods to exert parochial control.

The systemic reformulation of the social disorganization approach does not assume that the private and parochial levels of control only have effects on the level of neighborhood crime through the dynamics of affectively based expectations and supervisory capacities. In fact, to concentrate exclusively on these forms of community self-regulation would seriously distort the original argument of Shaw and McKay for, as Kornhauser (1978:38) has argued, they were also centrally concerned with the "effectiveness of socialization in preventing deviance." Shaw and McKay (1969:172) argued, for example, that children living in areas of low economic status are "exposed to a variety of contradictory standards and forms of behavior rather than to a relatively consistent and conventional pattern." Such a statement obviously indicates that there are subcultural aspects to the general Shaw and McKay model, and they eventually argued that certain neighborhoods were characterized by a "coherent system of values supporting delinquent acts" (1969:173); see Chapter 5 of this book). However, Kornhauser has observed that the cultural assumptions are not a necessary component of the social disorganization model. Rather, she argues that it is more consistent with the rest of their orientation to focus on the variability in the effectiveness of local structures of conventional socialization found in urban neighborhoods.

Some aspects of the socialization process entail processes of control at both the private and parochial level. For example, few criminologists would dispute the contention that the family represents one of the key socializing agencies in the United States. For example, in their recent discussion of low "self-control" as a key element of criminality, Michael Gottfredson and Travis Hirschi (1990:97–98) suggest that children are most likely to develop self-control when (1) their behavior is closely monitored, (2) deviant behavior is recognized by others when it occurs, and (3) such behavior is punished. Neighborhoods characterized by social disorganization are least likely to provide this type of setting for child rearing.

Sampson (1986, 1987a) has been perhaps most successful in incorporating family dynamics into the systemic neighborhood control approach to social disorganization, arguing that in relatively stable neighborhoods, parents often take on the responsibility for the behavior of youth other than their own children. However, in communities characterized by residential instability and heterogeneity and a high proportion of broken and/or single-parent families, the likelihood of effective socialization and supervision is reduced and it becomes difficult to link youths to the wider society through institutional means. Thus, cohesive family structures are effective sources of control "because they are aware of and intervene in group activities ... that are usually the predecessors of involvement in more serious delinquent activities" (1987a:107).

Unfortunately, other institutions of socialization have not received a great deal of attention in the social disorganization literature. This is most apparent in the general neglect of the educational process, even though it has been shown that rates of high school suspension are related to the neighborhood contexts in which the schools are embedded (see Hellman and Beaton 1986). The ethnographic work of Gary Schwartz (1987) suggests that the failure to consider the role of educational institutions within the larger context of the neighborhood may seriously limit our understanding of the processes of internal self-regulation. For example, he notes the example of Parsons Park, in which the local schools serve as "the cultural battleground of the community" (p. 50). In this neighborhood, it is primarily through the educational system that ethnic traditions, religious faith, family patterns, and other community standards are "woven into the fabric of local institutions." Yet in a second neighborhood located in the same city, the local high school "is experienced in the classroom as having little connection with the larger society's goals and values" (p. 222). It is no coincidence that these two areas also have very different rates of juvenile delinquency; the second, for example, has a heavy concentration of gang activity.

Although Shaw and McKay (1969) noted an association between rates of truancy and delinquency, they never fully developed an argument pertaining to the role of neighborhood schools. Neither did they discuss in any detail the role of local religious institutions, although Avery Guest and Barrett Lee (1987) have shown that a key church-related activity is the provision of social services aimed at the resolution of local community problems. Therefore, the role of local institutions (other than the family) as agents of control was relatively undeveloped by Shaw and McKay.

Perhaps the greatest shortcoming of the basic social disorganization model is the failure to consider the relational networks that pertain to the public sphere of control. The implications of this shortcoming have been illustrated in several studies that document the existence of stable neighborhoods with extensive interpersonal networks that nevertheless have relatively high rates of crime and delinquency (see, for example, Whyte 1981; Suttles 1968; Moore 1978; Horowitz 1983). The existence of such neighborhoods is an important contradiction of the predictions of traditional social disorganization approaches, which have focused only on the mutual linkages of residents. Efforts to explain this contradiction sometimes have led to the consideration of nonsystemic dynamics (often focusing on culture) that are logically inconsistent with the assumptions of a systemic model (see Shaw and McKay 1969 and Kornhauser's critique 1978).

However, as many urban analysts have noted (see, for example, Lewis and Salem 1986), it is very difficult to significantly affect the nature of neighborhood life through the efforts of local community organizations alone. Rather, these groups must be able to negotiate effectively with those agencies that make decisions relating to the investment of resources in the area that may foster the kinds of control that we have been discussing. For example, Whyte (1981:273) clearly argues that the problems he observed in the "Cornerville" section of Boston were due to the lack of effective ties between the neighborhood and the broader society; a similar argument is found in Suttles (1968). Therefore, a consideration of the public bases of systemic control is crucial to a full understanding of the relationship between neighborhood dynamics and crime.

The powerful potential for these external resources to foster a community's capacity for control is illustrated vividly in the history of the Conservative Vice Lords, one of Chicago's "supergangs" (for an excellent review of this history, see Dawley 1992). In 1968 and 1969, the Lords were able to solicit funds successfully from outside of their community (most notably from the Rockefeller and Ford Foundations) to develop a series of neighborhood-based programs. Although the central area of their territory generally was considered to be one of the most dangerous in Chicago, Dawley (1992) has argued that important changes in neighborhood life occurred during this period, including a significant decline in gang activity and a reduction in the fear of crime. Unfortunately, the Lords became involved in a series of intense clashes with City Hall and encountered problems with their tax-exempt status as an organization. As a result, the funding disappeared, and ten years later most of the Lords were either reinvolved in serious crime or dead, and the neighborhood regained its status as an "urban cemetery" (Dawley 1992:190).

Therefore, the existence of stable, high-crime neighborhoods in itself does not call the validity of the systemic model into question. Rather, it emphasizes the need to expand the focus of control beyond the internal dynamics of the community. We will examine recent extensions of the systemic approach that have attempted to address this issue later in this chapter. A summary of the basic systemic reformulation of the social disorganization model is presented in Figure 2–2.

Empirical Support for the Basic Systemic Model

Some serious empirical deficiencies have characterized most attempts to evaluate the social disorganization model. Traditional studies have used

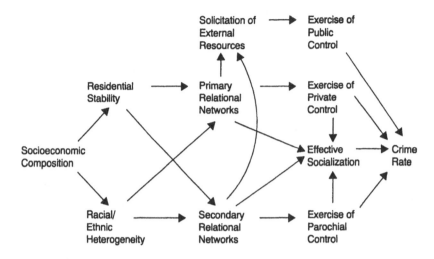

Figure 2–2. The Basic Systemic Model of Crime.

local community areas (Bursik and Webb 1982; Bursik 1984, 1986a, 1986b, 1989; Heitgerd and Bursik 1987), census tracts (or empirically based aggregations of those tracts into larger areas; Schuerman and Kobrin 1983a, 1983b, 1986; Shannon 1982, 1988, 1991; Taylor and Covington 1988), electoral wards (Sampson and Groves 1989), or police districts (Shannon 1988, 1991) within a given urban context as the units of analysis.[8] Some form of crime and/or delinquency data are then collected from a law enforcement agency, aggregated on the basis of the residential neighborhood of the offender, and the corresponding rates are computed. The compilation of these records in itself is a formidable task, but the information is usually housed in a central location and the costs of data collection (especially if the records are in a computerized form) are not high. Unfortunately, such research designs necessitate defining a neighborhood's boundaries on the basis of administrative jurisdictions. As we argued in the opening chapter, the selection of the appropriate spatial aggregation to be used as the unit of analysis in neighborhood studies is a complex issue with important implications for the findings derived from social disorganization research.

Most modern researchers are well aware of the problems inherent to the use of such boundaries. For example, while Bursik and Webb (1982) defend their use of the official local community areas of Chicago, they also note that it is a problematic approach. Likewise, Sampson and Groves (p. 783) note that the neighborhoods used in their analysis are only "approximations" of local communities. Nevertheless, most social disorganization researchers have resigned themselves to these restrictions on the determination of the neighborhood. However, there have been some attempts to aggregate data so that the units of analysis more adequately represent local communities. Schuerman and Kobrin (1983a), for example, clustered the 1,142 census tracts of Los Angeles into 192 neighborhoods whose homogeneity was determined empirically on the basis of physical contiguity, crime rates, land use, and demographic, socioeconomic, and subcultural features; ten of these clusters form the basis for Schuerman and Kobrin's research (1986). Such efforts are very similar in spirit to the methods used by Burgess and his colleagues to define the natural areas of Chicago over 50 years ago. Nevertheless, these approaches are neither able to address the symbolic issues implicit to the concept of neighborhood nor able to address the implications of the nested levels of neighborhood life that were discussed in Chapter 1. Therefore, without exception, the neighborhoods that have been used to test the social disorganization approach are less than optimally defined.

Even more important, these studies have been able to test only very limited aspects of the full social disorganization model. It is fairly easy to collect indicators of the ecological dynamics pertinent to the systemic approach (i.e., socioeconomic composition, residential instability, and population heterogeneity) from published census materials. In general, most studies have shown that the effects of these three ecological variables on the crime and/or delinquency rates are consistent with the predictions that can be drawn from Figure 2–2.[9] However, data pertaining to the nature of relational networks and systemic bases of control in the neighborhoods are not readily available from published sources. Thus, most large-scale studies of entire urban systems have been forced to assume that processes of systemic neighborhood control intervened between the ecological dynamics and the crime rate. Because of this lack of relevant data, many of the inferences that have been drawn from such research concerning the dynamics of social disorganization are based on "conjecture and speculation" (Sampson, 1987a:100). As might be expected, this state of affairs has compounded the confusion that exists concerning the measurement and conceptualization of social disorganization (see Byrne and Sampson 1986:13–17).

The acquisition of data pertaining to relational networks and the processes of control is generally only possible through an intensive series of interviews, surveys, and/or fieldwork within each of the local neighborhoods of a city. The logistic and economic costs of such studies in metropolitan areas, which may have over a hundred locally recognized communities, are obvious. One solution to this problem has been to restrict one's attention to the dynamics of systemic control within a limited number of neighborhoods, therby making the collection of relevant data feasible. Robert Kapsis (1976, 1978), for example, conducted a series of interviews with adults, adolescents, and community leaders who resided in three neighborhoods in the Richmond-Oakland (California) area. His results suggest that communities with broad networks of acquaintanceship and organizational activity have lower rates of delinquency, even when the racial and economic composition of the area would predict otherwise (see Table 2–1). The effects of such systemic control are most striking in his discussion of each neighborhood's response to a local racial disturbance that occurred in 1969.

Table 2–1
Characteristics of Neighborhoods in the Studies of Kapsis

	North Richmond	Iron Triangle	South Side
Total population	4,520	11,200	20,000
Percent black	93	40	80
Percentage increase in size of black population 1950–1965	3	36	64
Percentage increase in size of black population 1960–1965	2	6	28
Percent of household heads employed in skilled jobs	68	45	42
Percent of households with family income under $3,000	51	36	20
Percent of families on public aid	27	24	14
Unemployment rate	48	39	25
Percent of households with female head	27	13	16
Percent of homeowners	39	40	59
Percent of adults living in area less than 5 years	12	19	36
Grade point average of public school population (grades 7–12)	1.85	2.03	2.08
Percent of school population 16+ who dropped out of school 1965–1966	22	14	10
Official crime rate	0.92	0.99	1.21
Self-reported crime rate	0.74	0.81	0.76

Adapted from "Continuities in Delinquency and Riot Patterns in Black Residential Areas," by Robert E. Kapsis. © 1976 by the Society for the Study of Social Problems. Reprinted from *Social Problems*, Vol. 23, No. 5, June 1976, pp. 569–570 by permission of the author and publisher.

There were significant increases in law-violating behavior in the neighborhoods characterized by superficial and transitory relational networks, but the crime rate actually decreased in the most stable neighborhood as an indigenous citizen patrol committee emerged.

Kapsis (1978) cautions that his findings are drawn from a limited number of neighborhoods from one geographic region and that the full implications of the similarities and differences among those neighborhoods are not clear. Nevertheless, his studies do indicate that a neighborhood's capacity for systemic control may have an important mediating effect between the ecological composition of a community and its rate of crime and delinquency.

While this restricted reseach design is still relatively rare, two very important studies have appeared in recent years. As noted earlier in this chapter, Ora Simcha-Fagan and Joseph Schwartz (1986) have examined the delinquent behavior of a sample of youths residing in twelve neighborhoods in New York City. In addition to collecting traditional census materials for each neighborhood, they also used a survey to obtain data pertaining to attachments to institutions of conventional socialization (i.e., the school), the prevalence of family structures assumed to be related to effective socialization and supervision, the extent of informal neighboring, the extent of local personal ties, the level of sentimental attachment to the neighborhood, the size and breadth of local relational networks, and organizational involvement within the neighborhood. Thus, although they did not collect data concerning variation in the actual forms of private and parochial control found within the neighborhood (such as affectively based pressure or supervision), they were able to examine the capacities of each neighborhood to exert such control through relational networks.

The findings of Simcha-Fagan and Schwartz only partly support the predictions of the systemic approach to social disorganization. The residential stability and degree of organizational participation in the neighborhood were indirectly related to self-reported delinquency through their effects on attachment to the school; organizational participation also had a weak, although significant, direct effect on such delinquency. In general, however, not a great deal of support for the systemic model is provided by Simcha-Fagan and Schwartz.

Despite the major improvements that Simcha-Fagan and Schwartz made over the traditional social disorganization study design, their findings must be interpreted with some caution owing to the very small number of neighborhoods that are incorporated into their analysis. This is not the case in the recent study of Robert Sampson and Byron Groves (1989), who analyze data drawn from the 1982 British Crime Survey in which nearly 11,000 respondents residing in 238 electoral wards and polling districts were interviewed. Although the boundaries of these areas are administratively determined, Sampson and Groves (footnote 6) argue that they provide a good approximation to local communities. As we noted earlier in this chapter, we consider this study to represent the most complete examination of the systemic social disorganization model that has ever been attempted.

Unlike the mixture of survey and census data found in Simcha-Fagan and Schwartz, Sampson and Groves create all of their ecological systemic variables by aggregating the survey responses to the neighborhood level:

- Socioeconomic composition: the summed individual z-scores of the percent college educated, the percent in professional and managerial positions, and the percent with high incomes
- Residential stability: the percentage of residents brought up in the area within a fifteen-minute walk from home
- Racial/ethnic heterogeneity: an index of diversity based on the percent white, West Indian or African black, Pakistani or Bangladeshi Indian, other nonwhite, and mixed
- Local friendship relational networks: the average number of friends living within a fifteen-minute walk of the respondents' homes
- Organizational participation: the percentage of residents who had participated in meetings in the week before the interview
- Supervisory capacity: the percentage of residents who reported that disorderly teenage peer groups were a very common neighborhood problem

Thus, the dataset analyzed by Sampson and Groves contains material pertaining to the private and parochial control networks in the neighborhood and the extent to which these networks are able to control the activities of youths in the community, as well as indicators of the three commonly considered ecological dynamics. In addition, they also consider the effects of family disruption (the proportion of divorced and separated adults, and the percentage of households with single parents with children) and urbanization (a dummy variable reflecting a central city location).

The analysis of Sampson and Groves generally supports the predictions of the systemic social disorganization model (see Table 2–2). For example, extensive friendship networks are less prevalent in central-city neighborhoods and areas of high instability. Likewise, instability, heterogeneity, family disruption, and urbanization all increase the likelihood of unsupervised youth group activity within the neighborhood. Contrary to the predictions of the traditional model, the socioeconomic status of the neighborhood had a direct negative effect on the presence of unsupervised youth groups and a direct positive effect on the extent of organizational participation. Surprisingly, heterogeneity and instability were unrelated to this participation. Overall, therefore, their observed relationships between

Table 2–2

Regression Estimates of Effects of Community Structure on Dimensions of Social Disorganization in 238 British Local Communities

	Local friendship networks		Unsupervised peer groups		Organizational participation	
	B	t	B	t	B	t
Socioeconomic status	−0.06	−0.91	−0.34	−5.31	0.17	2.33
Ethnic hetero-geneity	0.02	0.34	0.13	2.04	−0.06	−0.83
Residential stability	0.42	6.35	0.12	1.90	−0.09	−1.26
Family disruption	−0.03	−0.45	0.22	3.73	−0.02	−0.28
Urbanization	−0.27	−3.91	0.15	2.21	−0.10	−1.32
R^2	0.26		0.30		0.07	

Adapted from "Community Structure and Crime: Testing Social-Disorganization Theory," by Robert J. Sampson and W. Byron Groves in the American Journal of Sociology, V94 (p. 788), 1989 by permission of the author and University of Chicago Press.

ecological change and neighborhood dynamics depart somewhat from those predicted by the traditional systemic model.

The central question is whether these dynamics are related to the likelihood of criminal behavior. Sampson and Groves focus on two alternative sets of crime indicators. The first reflects the rate at which the residents report that they have been victimized within the neighborhood and their perceptions of the rate of muggings and street robberies in the area. Although such variables are strongly related to the traditional concerns of social disorganization, they are conceptually distinct in that an unknown number of these victimizations are committed by nonresidents of the neighborhood. Therefore, the analysis of victimization rates entails a consideration of the neighborhood processes that are related to the probability that residents will be selected as the targets of crime. Since this is the central concern of Chapter 3, we will not discuss those issues at this point.

The second set of indicators provides a more direct operationalization of the local crime rate. Based on the self-reports of the respondents, Sampson and Groves created variables pertaining to the rate at which the residents engaged in personal violence and property crimes. From the perspective of social disorganization, the results are mixed (see Table 2–3). As would have been predicted from Figure 2–2, none of the three ecological variables have direct effects on either crime rate, but are mediated by the presence of unsupervised groups of teenagers (which increased the likelihood of both kinds of crime). Relatedly, given the argument of Sampson concerning family structure and the capacity for supervision, relatively high rates of family disruption make property crime more likely, although it is unrelated to personal crime. Although the extensiveness of friendship networks decreased the likelihood of property crime, it was unrelated to the rate of personal crimes. Finally, the rate of organizational participation was unrelated to either form of crime.

This research has several important implications for the basic systemic social control model that are worth highlighting. First, the systemic structure of a neighborhood seems to be more powerful in the control of property crime than personal crime. Second, the private sphere of community control (i.e., affective relational networks) seems to be more effective than the parochial sphere. Finally, the supervision of youths is a key component of effective neighborhood crime control.

Table 2–3
Regression Estimates of the Effects of Community Structure and Social Disorganization on Offending Rates in 238 British Local Communities

	Offending rates			
	Personal violence		Property	
	B	t	B	t
Socioeconomic status	−0.03	−0.04	0.40	0.54
Ethnic heterogeneity	0.14	1.79	−0.10	−1.27
Residential stability	−0.02	−0.30	0.10	1.24
Family disruption	−0.00	−0.08	0.18	2.56
Urbanization	−0.00	−0.11	−0.10	−1.23
Local friendship networks	0.02	0.20	−0.17	−2.26
Unsupervised peer groups	0.17	2.20	0.16	2.16
Organizational participation	0.01	0.16	0.08	1.27
R^2		0.06		0.09

Adapted from "Community Structure and Crime: Testing Social-Disorganization Theory," by Robert J. Sampson and W. Byron Groves in the American Journal of Sociology, V94 (p. 793), 1989, by permission of the author and University of Chicago Press.

Extensions of the Basic Model

As the preceding sections have indicated, a great deal of effort has been expended in recent work to clarify and test the systemic assumptions of the basic social disorganization model. As should have been apparent, however, much of this research has failed to address some of the key aspects of neighborhood control that were discussed in the opening chapter.

However, there have been efforts to expand the traditional scope of social disorganization beyond that discussed in the preceding section. These attempts have opened up new avenues of investigation that were either taken for granted by traditional disorganization theorists or were simply not considered. The findings have been extremely promising.

Disorder as a Mediating Condition of Systemic Control

For at least two decades, one of the central research issues of criminology has concerned the fear of crime. While we will extensively examine the role of the neighborhood in this context in Chapter 4, the notion of disorder that has developed within this literature has important relevance to the systemic control of criminal behavior (see Hunter 1978).

Disorder has been defined by Wesley Skogan (1990:4) as a violation of norms concerning public behavior; Dan Lewis and Greta Salem (1986:xiv) similarly refer to disorderly behaviors (or, in their terms, "incivilities") as reflections of the "erosion of commonly accepted standards and values." It has generally been used to refer to situations such as unacceptable behavior by teenagers, the physical deterioration of homes, commercial areas, or public spaces, the intrusion of "different" population groups into an area, or an increase in marginally criminal behaviors such as drug use or vandalism (see Lewis and Salem 1986:10). As such, it is a much broader normative conceptualization of problematic behavior than the definition of crime that we presented in Chapter 1 and is especially susceptible to charges that the criminologist is applying middle-class standards of behavior on populations that do not consider such behavior especially troublesome.

For example, there is a small commercial district directly across the street from the University of Oklahoma campus (referred to as Campus Corner) with a cluster of taverns, clubs, and restaurants that are very popular with the local and University communities. During the evenings, the streets of Campus Corner often are populated by small, but vocal congregations of people, many of whom are under the influence of legal and illegal substances. A relatively high proportion of these people are minors.

In the summer of 1990, the Norman Police Department formed a Street Crimes Unit, which was funded partly with federal money to focus on street-level narcotics users and narcotics-related offenses. On the evening of August

25, this Unit "swept" through Norman and made fifty-two arrests, targeting in particular the parking lots of restaurants and clubs in the Campus Corner area. Half of the arrests were for public intoxication (Cannon 1990).

Many residents of the general Normal community felt that perhaps this was an unwarranted display of force against a relatively minor problem. After all, according to the National Survey of Crime Severity (Wolfgang et al. 1985:47), the perceived seriousness of public drinking is extremely low (ranking 198th in a list of 204 offense descriptions). Yet the sweep was organized when the police department received complaints from residents and merchants in the Campus Corner area after a knife fight that occurred the preceding Friday. Therefore, attitudes were extremely mixed concerning the seriousness of the public behavior that was widespread in the area.

Lewis and Salem (1986:10) emphasize that what is seen as disorderly behavior may differ with the interests, values, and resources of neighborhoods, and they document a significant degree of variation in the perceptions of the seriousness of "incivilities" in their ten study areas. Skogan has discussed this issue extensively (1990:3–9) and has presented evidence that the problem of bias may not be especially severe. After controlling for the neighborhood of residence of the 12,813 respondents for which he had survey data, he examined the degree to which perceptions concerning the prevalance of disorder in the neighborhoods were related to individual characteristics such as race, age, education, and income (pp. 54–57). Although age was significantly related to such perceptions, he finds that there is a great deal of within-neighborhood agreement concerning the distribution of disorder: about 95 percent of the explainable variation was attributed to the neighborhood effect. Therefore, while there may be differences between neighborhoods in the perception of the level of disorder, there is solid empirical evidence that consensus exists within particular local communities.

The relationship of neighborhood disorder to crime is very important. James Q. Wilson and George Kelling (1982:31) have argued that disorder and crime are intertwined in a developmental sequence. Disorder can lead directly to increases in crime. Wilson and Kelling note that once symbols of disorder (such as broken windows or abandoned cars) become widespread in a community, behaviors such as vandalism are much more likely because disorder connotes the message that "no one cares." This rationale clearly was apparent in the justification for the street sweep given by the Norman Chief of Police: "We want to address problems before they get out of hand" (Cannon 1990:2).

However, disorder has a much more important indirect effect on crime in that it may lead to a breakdown of community control. As Skogan has argued (p. 47), when indicators of social disorder (such as public drinking, the presence of loitering youths and/or panhandlers, visible drug use, or sexual harassment) or physical disorder (such as abandoned buildings in disrepair or uncollected trash) become highly visible in a neighborhood, residents may feel demoralized, helpless, and angry at being crowded out of community life. In fact, Lewis and Salem (1986:74–76) note that residents were more concerned about disorderly behavior than crime in eight of the ten neighborhoods included in their study.

Skogan (Chapter 4) has documented the relationship between disorder and the systemic model of control that we have presented (see Figure 2–3). First, while the distribution of disorder is related to the level of poverty, racial composition, and degree of instability in a neighborhood, instability has the greatest effect. Second, high levels of disorder tend to be associated with lowered rates of mutual helping behavior among residents, satisfaction with the area, and stated plans to remain in the neighborhood. Thus, disorder can seriously disrupt the relational networks within a community, thereby decreasing its ability to control the behavior of its residents.

As noted in the beginning of this section, the focus on disorder arose within the fear of crime literature. It has also been incorporated into models of victimization, as will be discussed in the next chapter. Unfortunately, it is not yet a common element of most systemic models of criminal behavior despite work that highlights its relevance. For example, Sampson and Groves (1989) find that the visible presence of unsupervised groups of youths is significantly related to rates of personal and property crime. Likewise, Gottfredson and Taylor (1986:147–148) note that offenders with an extensive history of criminal involvement fare worse when released from prisons into neighborhoods ranking high on their incivilities scale. Such findings indicate that a consideration of the systemic implications of disorder should be a central focus of future research in this area.

The Effect of Changing Ecological Structures

The notions of change and adaptation are central to the theory of human ecology in which the social disorganization perspective is grounded. Thus,

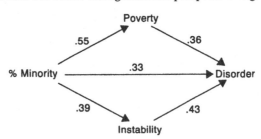

Figure 2–3. Ecological determinants of Disorder.

Adapted from *Disorder and Decline: Crime and the Spiral of Decay in American Neighborhoods* by Wesley Skogan. Copyright © 1990 by Wesley G. Skogan. Adapted from p. 75 with the permission of the Free Press, a Division of MacMillan, Inc.

the full set of dynamics that are related to a neighborhood's ability to control itself through relational networks can only be discerned when long-term processes of urban development are considered.

Shaw and McKay had access to a unique set of data that enabled them to analyze the relationship between ecological change and delinquency over a period of many decades. Without such a research strategy, it would have been impossible for them to document their important finding that local communities tend to retain their relative delinquency character despite changing racial and ethnic compositions (see the discussion of Stark 1987). However, the compilation of such information over an extended period of time is costly and often impossible. Therefore, with the important exception of the studies of Calvin Schmid (1960a, 1960b), most of the studies that appeared after Shaw and McKay's work were forced to rely on cross-sectional data.

This limitation had a major effect on the development of the social disorganization perspective, especially after Bernard Lander's cross-sectional study of Baltimore (1954) presented evidence that apparently contradicted Shaw and McKay's finding of a strong zero-order relationship between the socioeconomic composition of an area and its delinquency rate (see Bursik 1986a). Lander's research led to an important set of replications that confirmed the existence of the economic status and delinquency relationship (Bordua 1958–1959; Chilton 1964; Rosen and Turner 1967; Gordon 1967; Chilton and Dussich 1974). However, even after the issue was resolved, most subsequent studies continued to ignore the full dynamic implications of the social disorganization model and remained associational rather than processual formulations.

Such a limitation presents no problems if the ecological structure of an urban system is in a state of equilibrium. However, as Schuerman and Kobrin (1983a, 1986) have argued, the ecological stability assumed to exist by Shaw and McKay disappeared after World War II, when an acceleration in the rate of decentralization in urban areas significantly altered the character of urban change.[10] The effects of such developments on the distribution of crime and delinquency are impossible to detect without longitudinal data. Thus, the cross-sectional model of social disorganization were grounded in a basic assumption of stability that simply was not justified by historical evidence.

Recent work has been characterized by a renewed emphasis on the dynamics of ecological change in urban areas and their reflection in the changing distribution of crime and delinquency rates. Bursik (1986a), for example, has shown that the ecological structure of Chicago was relative stable between 1930 and 1940, concluding that at this time in the city's history, Shaw and McKay's arguments concerning the existence of ongoing areas of social disorganization and delinquency were generally upheld. However, this was not the case between 1940 and 1950, when important processes of ecological redefinition began to emerge related to rapid increases in the nonwhite composition of some neighborhoods. Thus, this period was a time of transformation of some neighborhoods into areas of relative instability. Since the formal and informal networks necessary to maintain effective self-regulation are difficult to maintain during periods of rapid compositional change, these areas also experienced concurrent increases in the rate of delinquency. Bursik and Webb (1982) and Bursik (1986a) argue that this was the primary reason why a moderate correlation between the racial composition of Chicago's neighborhoods and the delinquency rate emerged during this historical period.

Between 1940 and 1950, the most dominant form of rapid neighborhood change in Chicago was of the "foothold" variety, in which nonwhites increased their relative composition by at least 10 percent but remained in the minority in the community (Bursik and Webb 1982:38). This was not the case between 1950 and 1960, when most changing neighborhoods were characterized by the more rapid and dramatic shift of nonwhites from minority to majority status in the neighborhood. Some areas of the city experienced very rapid racial change during this period, such as the Kenwood neighborhood, which experienced an increase from 13.1 to 91.1 percent black during these 10 years. Because of the presumably more rapid disruption of local networks, the correlation between delinquency and ecological change was much stronger during this period than during the preceding decade. The process of neighborhood racial redefinition began to stabilize between 1960 and 1970 as nonwhites began to entrench themselves in their new neighborhoods. As McKay (1967:115) has shown, the delinquency rates in many of these newly black areas began to fall accordingly.

Schuerman and Kobrin (1983a, 1986) have provided a detailed examination of the sequences of ecological change involved in the transition of neighborhoods from low-crime to high-crime areas in Los Angeles. Over a twenty-year period, changes in land use patterns from predominantly owner-occupied dwellings to rental units led to changes in the population composition, population turnover, and socioeconomic composition of Los Angeles neighborhoods. The culmination of this process was a decrease in the prevailing controls in the area, which in turn increased the likelihood of crime and delinquency.

These findings seriously question the assumption of a stable ecological system and have two important implications for future developments in the social disorganization approach to crime and delinquency. First, it has already been noted that even at the cross-sectional level, the collection of data relevant to the dynamics of systemic control is very problematic. Since these dynamics have been shown to be affected by the ecological structure of the neighborhood, and since such structures in major cities have been undergoing significant changes during the last few decades, a full sense of the processes through which neighborhood dynamics are related to rates of crime and delinquency would necessitate a program of survey data collection at the local community level that spans at least several decades. Given the current funding situation in the social sciences, the prospects for such a study are fairly grim. Therefore, it is possible that we may never have the requisite data for a full test of the systemic approach and may have to resign ourselves to partial and incomplete examinations.

Second, the dynamics that gave rise to these ecological changes underscore the need to consider the systemic implications of the public sphere of control. In part, the changes that characterized Chicago after World War II were an outcome of the Supreme Court open housing decisions of 1948 that made restrictive covenants illegal (see Bursik and Webb 1982). However, Hirsch (1983:16) has argued that those decisions simply dealt a "final blow" to a practice that had not been effective in the protection of white homogeneity for many years. Rather, he attributes the rapid racial change during this period to the beginning of the white suburbanization process that followed the war, which provided a new source of available housing in the central city.

To some extent, the redefinition of many central-city neighborhoods caused by suburbanization was a result of "natural" ecological processes of growth. However, it is also true that much of the instability in these changing neighborhoods was caused by conscious decisions on the part of the housing and lending industries (see Hirsch 1983:Chapter 1; Bursik 1986a). The new housing construction that did occur shortly after World War II was of a type and in locations not suited to easy black resettlement. Taking advantage of this and the large increases in the black populations, landlords within the Black Belt of Chicago tended to subdivide existing apartments into much smaller "kitchenettes," many of which lacked bathroom facilities. Such efforts resulted in strong "push" factors that encouraged many residents to leave their neighborhoods even before it was economically feasible. Thus, because of their effects on the internal organization of many black neighborhoods, the actions of the housing and lending industries were at least partly responsible for the increase in the crime and delinquency rates that characterized these communities during this period.

These findings indicate that a full understanding of systemic control must also consider the effects of decisions and dynamics based outside of the neighborhood on the nature of relational networks within the community. Unfortunately, although Shaw and McKay were aware of such processes, this aspect of their work was largely undeveloped (see the criticism of Snodgrass 1976) and continues to be a shortcoming of most systemic approaches. However, recently there have been some important developments in this area.

The Public Level of Systemic Crime Control

As noted above, Shaw and McKay at least implicitly assumed that the ecological distribution and movement of populations within an urban area reflected the "natural" market of housing demand. Yet as Suttles (1972:41) has noted, modern neighborhoods reflect not only the economic processes considered by Shaw and McKay, but also "politics and some cultural image of what the city ought to be like." Thus, as Finestone (1976) has shown, the primary thrust of Shaw and McKay's model (as well as almost all subsequent work in that tradition) gives the impression that the composition and internal organization of local communities are relatively independent of the broader political and economic dynamics of the city (also see Bursik 1989). As a result, the public level of systemic control (Hunter 1985:233), which reflects the ability of a neighborhood to influence political and economic decision making and to acquire externally based goods and services that may increase its ability to control the level of crime in the area, has received almost no attention from social disorganization researchers.

Such an omission is simply not supported by work in the area of urban sociology. Recall that the human ecology framework assumes that as racial and ethnic groups are assimilated into the occupational structure of an urban area, they will be able to relocate into progressively more desirable residential neighborhoods. Therefore, as Theodore Hershberg and his colleagues (1979) have observed, the ecological structure of the city is the "material expression" of the occupational opportunity structure. Unfortunately, as industries have been economically enticed into suburban and rural locations, many residents of central-city neighborhoods no longer have easy access to the manufacturing jobs that traditionally provided the opportunity for occupational mobility among urban immigrant groups, especially in the older, Northern cities. Hershberg et al. (p. 73), for example, note that Philadelphia lost 75,000 manufacturing jobs between 1930 and 1970. Likewise, between 1954 and 1982, Chicago experienced a decrease from 10,288 to 5,203 manufacturing establishments, and the number of workers they employed fell from 615,700 to 277,000.

Changes in the political economies of urban areas should be a central consideration in systemic models of neighborhood crime, for the stability of neighborhoods depends in part on the stability of jobs (see Hershberg et al., p. 78). Loic Wacquant and William Wilson (1989) note that the exodus of manufacturing jobs from central cities and the accompanying shift toward service industries and occupations have resulted in the social and economic marginalization of blacks and the deterioration of predominantly black neighborhoods. Such developments can have devastating effects on the ability of local communities to act as agents of social control. For example, Wacquant and Wilson provide data (pp. 22–24) suggesting that there has been a decline in attachment to and identification with the neighborhood, fewer social ties with other community residents, and a loss of strength in local community organizations. Barry Bluestone and Bennett Harrison (1982) also note that the closing of manufacturing plants is typically accompanied by strained family and social relationships and a general decline in social cohesion.

A focus on the public level of crime control would examine how neighborhoods have lobbied industry and government in an effort to keep such jobs available to neighborhood residents. A relatively large literature has emerged on the reactions of local communities to the impending closing or relocation of industrial plants (see, for example, Bluestone and Harrison 1982, Buss and Redburn 1983, or Portz 1990). While the full implications of such community dynamics have yet to be incorporated into systemic models of crime control, there have been some attempts to examine one of the outcomes of this process—the emergence of the black underclass.

Most notable in this regard has been the work of Ralph Taylor and Jeanette Covington (1988), who argue that the increasing social isolation of the urban underclass leads to a heightened sense of relative deprivation, which in turn should increase the likelihood of violent crimes. They test this proposition using 1970 and 1980 neighborhood-level data collected in Baltimore. Overall, they find that both relative deprivation and the more traditional variables used in social disorganization studies are related to violent crime. In the underclass neighborhoods, for example, relative deprivation was more strongly related to murder rates than the stability of the neighborhood; however, the reverse was true for crimes of assault. In addition, they find that the relative effects of these two sets of variables depend on the economic composition of the neighborhood. For example, evidence is presented that the social disorganization model is more powerful than relative deprivation in explaining changes in the rates of violent crimes in neighborhoods experiencing rapid economic improvement through the process of gentrification. Such results suggest that modern urban dynamics related to the changing political economies of American cities have significantly modified the role of economic deprivation as a cause of neighborhood crime rates. Future investigations of this issue should lead to some important improvements in the systemic model.

Although the occupational opportunity structure found in modern metropolitan areas is one arena in which neighborhoods may attempt to exert public control, others are equally pertinent to a systemic approach to crime. Just as the traditional human ecology model assumed that the occupational opportunity structure was relatively open to all immigrant groups, it also assumed that the extent to which housing was available to such groups was primarily determined by the dynamics of the free market. As we discussed in Chapter 1, this has not been the case since World War II with the development of public bureaucracies mandated to determine the allocation and use of land. The decisions of such agencies can dramatically affect the stability of neighborhoods through demolition and construction, disinvestment of city resources, and the creation of social climates conducive to real estate panics (see Skogan 1986:206–207).

There are many situations in which neighborhoods may attempt to influence decisions made concerning the use of land within its boundaries. Manuel Castells (1983), for example, has provided a very detailed history of the Mission Coalition Organization, which grew out of an organized (and successful) effort of community residents to resist the potential demolition and displacement of the Mission neighborhood of San Francisco that would have resulted from a proposed urban renewal program (for additional examples of community resistance to externally imposed changes, see Feagin and Parker 1990). Unfortunately, many attempts of local communities to exercise public control of their neighborhoods have not been as successful. Residents of Harlem and other areas of New York City have recently been expressing a great deal of concern over the fact that their neighborhoods have the highest concentration of housing and related programs for the homeless, the mentally ill, and drug abusers while other sections of the city have very few (or none) of these programs (see Chira 1989). While the city officials have expressed an awareness of this problem, they have argued that the situation has arisen primarily due to economic considerations. However, the residents have offered the counterargument that these neighborhoods have been selected because they are relatively powerless to resist their imposition. In the terms of this book, such sentiments reflect the inability of existing relational networks to exert any effective form of public control.

As was the case with the political economy, there have been very few attempts to examine how such political dynamics might be reflected in the crime rates of the affected areas. One exception is Bursik's (1989) study of public housing project construction that was noted in the opening chapter. This controversial type of planned community change has resulted in some pronounced conflicts between residents and the local government. To a large extent, this conflict reflects the general public image that such housing is primarily designed for slum clearance and is usually designated for location in rundown areas (see Foley 1973). Therefore, residents either may feel that they have been abandoned by the local government or may perceive that the future of the community has been greatly jeopardized.

Three programs were developed under the 1974 Housing and Community Development Act to provide subsidized housing for families who earned up to 80 percent of the median income in a particular city (see Weicher 1980). In one of these, the new construction program, subsidies were given to particular projects rather than to individual families. It is important to emphasize that the cities incurred no cost for the construction of new dwelling units, even though the cost per unit rates were significantly higher than for the subsidy of units in the other two programs. In light of the fiscal crises faced by many cities because of the imbalance between the financial obligations of the municipal governments and the taxes drawn from the private sector, the potential benefits that might be derived from new construction are obvious: the channeling of funds to local developers, the employment of large numbers of construction workers, and an increase in the local tax base.

Decisions have to be made concerning the placement of such projects. If a strictly open-market housing mechanism prevailed, one would expect that these units would be built in areas in which the costs to the program were minimal, in terms of both the value of the land that had to be purchased and the fair market rents that the federal government would be willing to subsidize. This assumption is identical in spirit to the justifications that were made by the New York officials concerning the concentration of social service programs in only a limited number of neighborhoods. On the other hand, if in fact these locational decisions reflected the outcome of conflict-based political decisions, one would expect that the neighborhoods least able to organize an effective resistance to their construction would be the primary targets.

Bursik (1989:115–116) presents evidence that the public housing built in Chicago under this program was most likely to be constructed in areas characterized by high rates of residential instability. In addition, the market value of the occupied housing units in the neighborhood, which was used to represent the costs involved in the purchase of land or demolition of existing units, was not related to the likelihood that public housing would be erected in Chicago's local communities. This finding is extremely significant, for it indicates that the dynamics typically associated with the existence of an open, competitive housing market were not related to these political decisions. Rather, neighborhoods characterized by such construction tended to be those least able to exert effective public control over these events.

While the construction of public housing did not lead to any significant changes in the racial and economic composition of Chicago's neighborhoods, there was a marked association with the subsequent degree of instability in these areas: on the average, neighborhoods experiencing such construction were characterized by an 8 percent increase in instability in 1980 over what would have been expected on the basis of the 1970 patterns. Given the significant effect of instability on the delinquency rates of Chicago's neighborhoods (p. 116), such findings indicate that the political decision making associated with the construction of such projects indirectly accelerated existing rates of delinquency.

Other characteristics of the housing market that may be related to the level of crime and delinquency found within a neighborhood also fall within the domain of the public order of control. James Burnell (1988) has shown that when an individual or family makes decisions about a potential residential location, the characteristics of that particular neighborhood are not the only aspects of the location that are taken into consideration. On the basis of his findings that the rate of crime and racial/ethnic composition of neighboring communities also play an important role in such decisions, Burnell argues that potential residents are concerned about the possibilities of crime spillover (see Hakim and Rengert 1981) and the effects on future property values.

As Suttles notes (1972:201), many adults in communities in which there is a perceived threat of unwanted racial invasion from nearby neighborhoods may support juveniles in illegal activities that are considered by the adults to represent a degree of protection of life and property from the "dangerous" residents of adjacent communities. Such dynamics are a hallmark of what he has referred to as the "defended community" (1968, 1972). This position is strongly supported by the work of Hirsch, who has documented the "communal" aspects of most of Chicago's housing riots that occurred between 1947 and 1957 (1983:Chapter 3). For example, he notes that while most of the persons arrested during these riots were young, they were fully supported by older men and women. Therefore, the "violent ones who were most often arrested were the community's representatives, not . . . its aberrations" (1983:74).

When the 1984 paper of Bursik was first submitted for review, an astute reviewer (whom we later learned was Roland Chilton) noted that some of the empirical patterns were so unusual that they might be caused by the presence of an outlier in the data. Upon examination of this possibility, the dynamics found in one particular Chicago neighborhood called Burnside did, in fact, appear to have influenced significantly the aggregate relationship during 1950 (see Bursik 1984). At this time, Burnside was a small, sharply defined enclave of Hungarians and Poles. In fact, its concentration of foreign-born whites was the second highest in the city (25.6 percent); only 0.2 percent of the population were nonwhite. It was also a predominantly blue-collar community (only 2.2 percent of the labor force were professionals) with a great deal of residential stability (60.7 percent of the dwelling units were owner-occupied). Therefore, any criminologist even vaguely familiar with the social disorganization tradition would have predicted that it would have a fairly low rate of crime and delinquency. However, during 1950, it had the highest delinquency rate in the city of Chicago—47.2 court referrals for every 1,000 males aged 10 to 17. This was almost twice as high as the second most delinquent area in Chicago (24.1) and far greater than the average rate found in the other local communities (6.8).

Part of this large rate may be accounted for by the small population of Burnside—there were only ten court petitions filed for 212 male youths. However, Burnside was also in a position to exhibit some of the characteristics of the defended community. In 1949, two riots involving the movement of blacks into previously all-white areas (one of which received national exposure) took place in neighborhoods that were not far away. In addition, the community immediately to the southwest had changed from 4.2 percent to 18.4 percent black between 1940 and 1950. Burnside residents may also have foreseen future developments in the neighborhood that surrounds it to the west and north—it changed from 0.8 percent to 63.7 percent black between 1950 and 1960 while Burnside maintained its nearly all-white status. Thus, actual and potential racial changes were occurring on all sides of it.

Although Bursik (1984:408) proposed that the high rate of delinquency in Burnside represented community efforts to control dynamics taking place outside of the community, he did not test that hypothesis. This possibility was looked at extensively by Janet Heitgerd and Bursik (1987). Their most important finding was that even after controlling for changes that occurred between 1960 and 1970 in the household characteristics, racial/ethnic composition, and stability of Chicago's neighborhoods, the rate of racial change in adjacent neighborhoods was significantly related to increases in the rate at which youths living in the community were referred to juvenile court.

It is important to emphasize that this evidence concerning the relationship between the public order and the control of crime is indirect since, like all traditional datasets, information is not available pertaining to the central systemic dynamics assumed to underlie the association. However, Heitgerd and Bursik (p. 784) provide additional support for their argument by noting that in the least stable neighborhoods (in which it is assumed that the controlling capacities are weakest), compositional changes in neighboring areas were not significantly related to increases in delinquency. That is, these communities did not appear to have been in a position to encourage delinquency as part of an organized attempt to control these external dynamics. On the other hand, there was a very strong association within the most stable neighborhoods.

The findings of the studies that have been discussed in this section all suggest that the inability to address issues of public control has been a significant shortcoming of the traditional systemic approaches and has led to a seriously incomplete understanding of the neighborhood dynamics related to crime and delinquency. However, they also indicate that such research is possible on at least a limited basis.[11]

The Effect of Crime on Systemic Forms of Control

The research discussed in this chapter has generally focused on the extent to which rates of crime and delinquency depend on the ability of local communities to regulate themselves. It may be, however, that most systemic models are substantively incomplete because they fail to consider the degree to which rates of crime and delinquency may also affect a community's capacity for social control.

The rationale for the consideration of such reciprocal relationships has been developed most thoroughly by Wesley Skogan (1986, 1990) in his discussions of neighborhood feedback loops. As Skogan (1986) argues, the level of crime has a marked (although imperfect) effect on the fear of crime experienced by residents of that area. High levels of fear, in turn, may result in the following:

1. Physical and psychological withdrawal from community life
2. A weakening of the informal social control processes that inhibit crime and disorder
3. A decline in the organizational life and mobilization capacity of the neighborhood
4. Deteriorating business conditions
5. The importation and domestic production of crime and delinquency
6. Further dramatic changes in the composition of the population

Thus, low levels of systemic control increase the likelihood of crime, high levels of crime decrease the effectiveness of systemic control, and the entire process spirals onward.

Such dynamics of the systemic model have not been completely ignored in the literature. Lyle Shannon (1988, 1991), for example, finds that there was a great deal of variation in the manner in which the relationship between ecological change and crime/delinquency rates have unfolded in Racine, Wisconsin. Nevertheless, he does conclude that there was a general process in which neighborhood deterioration and movement out of an area was followed by increase in the crime and delinquency rates, which in turn led to a further deterioration of the local community. Similar conclusions are reached by Schuerman and Kobrin (1986), who conclude that increases in the crime rate are followed by shifts in local land uses, population and socioeconomic composition, and normative structures.

Bursik (1986b) has focused in particular on the effect of increases in the delinquency rate on the racial/ethnic composition of neighborhoods and presents evidence that changes in this composition in Chicago's neighborhoods between 1960 and 1970 (which primarily reflected increases in the proportion of the population who were black) were significantly related to simultaneous increases in the delinquency rate. However, the magnitude of this effect was not nearly so great as that for the effect of changes in the delinquency rates on concurrent changes in the racial composition of these areas. Such findings suggest that a large part of the traditionally high association between race and crime may reflect situations in which minority group members are stranded in high-crime areas from which they cannot afford to leave.

Given the very few models that have examined the effect of crime on ecological structures and systemic forms of control, such results should only be accepted tentatively. A great deal of work must be done concerning the identification of these models, appropriate methods of estimation, and so forth. But they do represent the source of potentially important revisions of the systemic model.

Conclusion

Of all the issues to be discussed in this book, those that have been addressed in this chapter easily have the longest tradition of research. In general, the predictions of the systemic model have received support in a wide number of contexts. Nevertheless, it should be obvious to the reader that, because of serious data-related deficiencies, we still have a very incomplete empirical documentation of how neighborhood processes evolve to make criminal behavior of residents more or less likely.

Despite these limitations, the important modifications of the traditional social disorganization model that have been discussed indicate that it is not a hoary old framework with little relevance to modern criminology. Rather, the research that has been generated suggests that it is easily adaptable to modern development in urban sociology. Thus, the future of this tradition appears to be bright and exciting.

3

Neighborhood Opportunities for Criminal Behavior

> Man . . . life's a risk. But I can tell you this . . . when I'm cruising,
> looking for a hit . . . there are parts of town I won't touch.
> —Convicted property offender quoted in Carter and Hill (1979:47)

Chapter 2 focused on neighborhood processes that affect the probability that residents engage in illegal behavior. In this chapter, we
shift our attention to dynamics related to the distribution of opportunities for criminal behavior within a neighborhood. There is certainly
a great deal of overlap in these two orientations, for as we noted in Chapter
2, offenders tend to commit a sizable amount of crime within their own
local communities. Nevertheless, a large proportion of the crimes that occur
within an area are committed by people who reside elsewhere (and for
property offenses it is typically a majority of the crimes; see Rand 1986).

Since the late 1970s, a large literature has examined the ecological
processes that give rise to the spatial distribution of situations conducive to
the commission of crime. Since the orientation of this work emphasizes the
neighborhood distribution of criminal opportunities rather than criminal
populations, somewhat different dynamics are implicated than those we
discussed in Chapter 2. Nevertheless, these approaches are fully congruent
with a systemic theory of community control.

Neighborhood Selection as a Decision-Making Process

Paul and Patricia Brantingham (1984:Chapter 12) have provided a useful
description of the dynamics through which a neighborhood may be chosen
as a target for criminal behavior. First, it is obvious that potential offenders
cannot commit crimes unless they are aware of areas of the city that provide
likely opportunities for such activities; these areas comprise the "awareness
space" of the criminal. Therefore, it should be no surprise that familiarity
with a neighborhood is a central consideration in the decision to commit a

crime in a particular part of town (see Brantingham and Brantingham 1984; Carter and Hill 1979; Rengert 1989). In fact, criminals rarely travel into "unknown" neighborhoods to engage in illegal activity (Reppetto 1974). Rather, they appear to restrict their activities to those areas of the city in which they have been able to make observations and gather information through their non-crime-related recreational and work travel patterns (see Rengert and Wasilchick 1985:63–72).

Since profitable targets of crime are not randomly distributed throughout the awareness space of the potential criminal, the next stage involves an evaluation of those areas within the "awareness space" that are above a threshold expectation of profitability and probability of success. Brantingham and Brantingham (1984:363) refer to this as the "crime occurrence space." For example, Rengert and Wasilchick note (p. 62) that all their respondents were especially afraid of the consequences of breaking into a house occupied by a member of the Pagans, a local motorcycle gang. Therefore, the two areas of the city most heavily populated by the Pagans were not considered to be optimal neighborhoods for their acts of burglary.

Finally, the potential offender chooses an area from within the crime occurrence space that he or she considers to be most suitable for criminal activity. It generally has been shown that the affluence of the neighborhood is used by an offender as an indicator of the expected gain to be made in a property offense (see Cromwell et al. 1991:33). However, there does not appear to be widespread agreement about how such information is used. For example, Rengert and Wasilchick (p. 13) present evidence that burglars prefer areas of midpriced housing just outside their residential areas. Carter and Hill (1979), on the other hand, suggest that black and white property offenders tend to chose areas of relatively low socioeconomic status.

The conceptual scheme of the Brantinghams has been criticized for the implication that target neighborhoods are chosen on the basis of a highly rational calculation of costs and benefits (see Cromwell et al. 1991:10). Several criminologists (such as Cromwell et al. 1991; Rengert and Wasilchick 1985) have offered a counterperspective based on an assumption of "limited rationality." While such an orientation is consistent with a rational choice approach to criminal decision making (see Cromwell et al.:11), it argues that most illegal activities are the result of a simplified, immediate evaluation of the short-term costs and benefits of engaging in criminal behavior. For example, 35 percent of Bennett and Wright's sample of convicted burglars reported that they had not been worried about getting caught and sentenced, and another 26 reported that they had not even thought about it (1984b:187). Likewise, Cromwell et al. (1991:33) find that the key initial evaluation of a criminal opportunity by burglars is simply an intuitive sense of whether the potential gains of the burglary exceed the apparent risks.

By extension, the limited opportunity approach would argue that it is rare (and generally impossible) for potential offenders to calculate the potential gains to be made in each of the areas comprising their awareness space. Rather, these decisions are most typically made very quickly in the context of the situations that present themselves during the course of daily activities. For example, David Downes (1966:203) notes the justification he was given for a joy-riding incident: "The bloke just left it [a set of keys] lying around [at work]. Well, that's just asking for it, ain't it?" This spontaneity could account for the observed general clustering of illegal behaviors within the vicinity of the offender's residence when in fact the offender may be aware of areas of the city with more profitable possibilities.

While we are persuaded by the arguments of the limited rationality model concerning criminal decision making, very little is actually known concerning the processes by which potential offenders select target neighborhoods from within their "awareness spaces" (Rengert 1989).[1] One reason for this relative dearth of research is that it is difficult to collect reliable data concerning the decision-making processes that lead to the selection of particular neighborhoods. For example, none of the respondents in the Carter and Hill study could clearly state what it was about a particular community that influenced the decision to commit crimes in that area (p. 49). However, it is possible to approximate the decision-making processes that lead to the selection of certain areas of the city for criminal activities by comparing the characteristics of neighborhoods with varying rates of victimization (see Rengert 1989:168). All the research that will be discussed in this chapter has been conducted in such a manner.

The Emergence of Opportunity Theories of Criminal Victimization

It may seem fairly trivial to note that a potential target and offender must converge in time and space for a crime to occur. Yet this definitional aspect of crime was rarely incorporated into major criminological theories until relatively recently. For example, Shaw et al. (1929) noted that commercial and industrial areas that provided the opportunities for property crime were prominent features of neighborhoods with high residential delinquency rates. However, the ecological dynamics that spatially distributed these criminal opportunities do not appear in the Shaw and McKay model of social disorganization. Three decades later, Terence Morris (1957:20–21) still felt it necessary to argue that it was important to differentiate between the background characteristics that may predispose an individual to crime and the situations in which these potentialities become actualized.

A major breakthrough occurred in 1965 when Sarah Boggs shifted the focus of her research from the traditional residential neighborhood of the offender to the neighborhood in which the crime was committed. On the basis of the rates she computed for the census tracts of St. Louis, Boggs concluded that the selection of a neighborhood for criminal activities was a function of the familiarity between offenders and their targets, and the potential profitability of the criminal act. Several years later, Leroy Gould (1969) reached similar conclusions.

The publication of two important and popular books in the early 1970s greatly accelerated the growing interest in targets of crime or, as they generally came to be known, criminal opportunities. The first, *Defensible Space* by Oscar Newman (1972), was ostensibly concerned with architectural design. However, Newman's primary emphasis was on "the real and symbolic barriers, strongly defined areas of influence and improved opportunities for surveillance . . . that combine to bring an environment under the control of its residents" (p. 3). That is, Newman's thesis was grounded in the assumption that particular designs could decrease the likelihood of successful criminal victimization. Two years later, Thomas Reppetto's *Residential Crime* (1974) stressed similar themes, such as physical design, the visibility of potential crime sites, and the travel and work patterns of potential victims.

An important related development within criminology was the appearance of three major, federally funded studies (Biderman et al., 1967; Ennis 1967; Reiss 1967) that focused, in part, on the rates at which particular segments of society were victimized by criminal behavior. These works appeared at about the same time that feminist, law and order, and civil rights movements were expressing deep concerns for the protection of the rights of victims of crime (see Karmen, 1984), and it became increasingly clear that the dynamics through which individuals and groups provided opportunities for victimization (either or themselves of their property) should be an essential component of theories of crime (see Gould 1969:50). The growing emphasis on victimization research culminated in the publication of *Victims of Personal Crime* by Michael Hindelang and his colleagues in 1978, in which the first major theory of victimization was developed. This perspective (called the "life-style model") assumes that the manner in which individuals allocate their time to vocational and leisure activities is related to the probability of being in a particular place at a particular time. Since these life-style differences affect the amount of time that one spends in contact with people who are likely to commit crime, they affect the likelihood that someone (or someone's property) is exposed to situations in which there is a high risk of victimization (Gottfredson 1981: 720; Hindelang et al. 1978; Garafalo 1987).

Several aspects of the life-style model are worth emphasizing. First, it emphasizes the factors that make the convergence of a target and an offender within a particular neighborhood likely at a particular time (Garafalo 1987:26). Second, it takes for granted that a population of motivated offenders exists within an urban area and does not address the sources of that motivation. Therefore, it holds constant the offender-based factors that may lead to crime to focus on how the opportunities presented by potential victims in a particular situation (for our purposes, a neighborhood) may elicit a criminal activity. Finally, it emphasizes the patterned life-styles of population aggregates rather than variability in individual characteristics that may elicit victimization (Garafalo 1987:27).

Unfortunately, the dataset that Hindelang et al. used to develop their life-style model did not contain any information pertaining to the neighborhood contexts of victimization. Although they suggest in several places that neighborhood dynamics may be intimately connected to the patterns of victimization that they observed (see, for example, pp. 119, 148–149), it was not possible for them to directly address such a proposition. Therefore, the relationship of community processes to victimization was largely undeveloped within the original life-style model.

The Routine Activities Approach

At the same time that the life-style model was being developed, a similar perspective was being formulated by Lawrence Cohen, Marcus Felson, and their colleagues (1979, 1980, 1981), which has come to be known as the "routine activities" approach. The concept of a routine activity is nearly identical to that of a life-style, i.e., the patterned allocation of time to occupational and leisure activities, and some have argued that the differences between the perspectives are more apparent than real (see Maxfield 1987).[2] However, the two approaches differ somewhat in terms of the processes that are assumed to give rise to these activities. While Hindelang et al. (1978:242–244) argue that life-styles arise as individuals adapt to the structural constraints and role expectations imposed on them, Cohen and Felson assume that routine activities reflect the temporal and spatial distributions of the key sustenance activities of a community, such as the search for "property, personal safety, sexual outlet, physical control, and in some cases even survival itself." Crime, therefore, represents a struggle for these "goods" and occurs when the sustenance activities of a motivated offender and a suitable target converge in a particular location in the absence of a guardian that is capable of preventing the violation (Felson and Cohen, 1980:392–393).

The urban dynamics related to the offender/target/capable guardian convergence in a particular community are central to the routine activities approach. In addition, although the model has a very different emphasis than the social disorganization perspective, both frameworks historically are grounded in the human ecology tradition and share a clear systemic orientation to the neighborhood control of crime. Thus, the routine activities and social disorganization approaches provide complementary frameworks for the study of neighborhoods and crime.

The Ecological Basis of the Routine Activities Model

In 1944, Amos Hawley published an influential paper in which he criticized the traditional sociological formulation of human ecology. While he acknowledged that Park and Burgess's emphasis on emergent spatial patterns should be part of a general ecological model, he argued that their restricted focus led to a failure to come to grips with the central problem of the discipline: the development and the form of community structure in varying environmental contexts (Hawley 1944:404). Therefore, according to Hawley, the proper focus of human ecology is the overall structure of functionally interdependent relationships through which a localized population derives sustenance from the environment (Hawley 1950:180). Within this broader perspective, spatial distributions are ecologically significant only insofar as they help specify the nature of such relationships (1950:178).

These dependent relationships take two basic organizational forms. Symbiotic relationships represent the mutual dependence among functionally different individuals or groups that place different demands on the environment (1950:36–37). These relationships form the basis of internally differentiated, yet integrated corporate units in which the activities performed within the group supplement and complement one another in order to produce goods and services (p. 210).[3] It is the cluster of symbiotic groups within a community that mediates between the population and the environment from which it must draw its sustenance.

The second form of ecological organization is called commensalistic and represents collections of functionally homogenous individuals who make similar demands on the environment (1950:39); Hawley refers to these as categoric groups (p. 210). Because of their similarities, such groups can only engage in simple forms of collective actions, are characteristically reactive, and their primary function is "to conserve or protect what is necessary to the welfare of its members" (p. 211).

Hawley considers the modern city to be the "consummate example of the corporate unit" in which none of the groups that exist within the city can maintain themselves without the aid of the others (pp. 215–216). Thus, cities reflect a very complex set of functional interdependencies that Hawley refers to as the "web of life." Given this orientation, the key question for our purposes concerns the role of the residential neighborhood within this system of relationships. In this respect, his model is very similar to that of Park and Burgess. As Hawley notes (1950:278–282), family units are distributed with regard to land values, the locations of other types of units, and the time and cost of travel to centers of activity within the larger community. The key indicator of these three distributional dimensions is residential rental values (p. 281), which generally tend to increase with distance from the central city and give rise to the concentric zone patterns discussed by Burgess.

Hawley further argues (p. 282) that the presence of a given type of family unit is typically a "localizing factor" for other similar family units. This leads to the clustering of families with similar incomes into a relatively few neighborhoods within the urban area. However, income is not the only factor that may give rise to the clustering of family units, for they also can converge spatially on the basis of other shared similiarities, such as racial or ethnic status. If enough families share a spatial locality within the urban system, it becomes possible to "attract special services to their area . . . engage in their own peculiar forms of collective behavior . . . and offer relatively effective opposition to undesirable encroachments from without" (p. 282).

Thus, the systemic social control implications of the Hawley model are very similar to those of the social disorganization model that were derived from the Park and Burgess model in Chapter 2. Neighborhoods can be viewed as a particular type of symbiotic corporate unit, characterized by a division of labor and function among residents. Yet despite the differentiation that exists within the local community, the population of that area shares a common residential identification. Hawley's thesis implies that such a commonality can serve as the basis of categoric (i.e., commensalistic) behavior when the neighborhood is threatened. In such situations, the community may "coalesce in a united action to protect the common possession . . . internal conflicts and differences are set aside as the entire population rises to meet the threat" (1950:219).

As Felson (1986:123) has argued, "a tight community—where people know people, property, and their linkages—offers little opportunity for common exploitative crime." In this respect, Felson notes that the routine activities approach is compatible with Hirschi's (1969) theory of social control, which is simply a social disorganization model framed at the individual level. Therefore, the "capable guardian" component of the routine activity model essentially is identical to the broader concept of systemic control in the social disorganization framework. This equivalence is evident in the indicators of capable guardianship suggested by Massey et al. (1989:386): the relative proportion of one's day spent in and around the household, the number of people living in the home, the general spatial propinquity of other primary group members, the general level of protectiveness exhibited by neighbors, and the willingness to utilize formal and informal mechanisms of control.

One of the most interesting arguments of Hawley's ecological framework is reflected in the routine activity definition of crime as "predatory behavior," i.e., illegal acts that involve direct physical contact between at least one offender and at least one person or object in which someone definitely and intentionally takes or damages the person or property of another (Glaser 1971:4; Cohen and Felson 1979:589). As Felson and Cohen point out (1980:391), Hawley considers predatory behavior to be "a special case" of symbiosis, since it involves mutual relationships between the predator and the prey that affect the sustenance of both. While Hawley notes that parasites are sometimes considered unique because they draw their sustenance from other organisms in the community "without paying their way" (1950:38), he clearly rejects this position, quoting extensively from the animal ecologist Charles Elton:

> It is common to find parasites referred to as if they were in some way more morally oblique in their habits than other animals, as if they were taking some unfair and mean advantage of their hosts.... [W]e find that the whole animal kingdom lives on the spare energy of other species or upon plants, while the latter depend upon the radiant energy of the sun. If parasites are to occupy a special place in this scheme we must, to be consistent, accuse cows of pretty larcency against grass and cactuses of cruelty to the sun (Elton 1927:75).

Within the context of the routine activities framework, therefore, criminals are considered to be different from noncriminals only to the extent that they have developed alternative techniques of gaining economic and social sustenance from their environment. Thus, Cohen and Felson (1979:590) assume that there are two "functionally dissimilar" general populations in a community, one of which derives its sustenance through behaviors that have been defined as legal, and one of which derives its sustenance by taking resources from the law-abiding population. It does not attempt to account for the presence of these two groups and simply takes their existence as given.

The assumption of these two relatively distinct populations within an ecological system represents a rather dramatic simplification of Hawley's theory, which states that since symbiotic and commensalistic groups may interpenetrate one another at numerous points, "every individual may be thought of as standing at one or more intersections of the symbiotic and commensalistic axes" (1950:210). Our interpretation of this statement is that people may have multiple sustenance relationships with the environment; i.e., a person may represent an opportunity for victimization at the same time that he or she is obtaining sustenance from illegal means. Such a conclusion receives strong support from the research of Singer (1987:171–179), who finds that gang members are more likely to be victims of violent offenses than nongang members and that individuals who use weapons to threaten other people are more likely to be the victims of property and violent offenses than those who do not use such weapons (see also Sampson and Lauritsen 1990). In this respect, the dual-population assumption of the routine activities approach makes it much more limited than the social disorganization model, which argues that a community resident can be either an offender or nonoffender depending on the nature of the systemic controls that are present in particular situations.

If this short description represented the full implications of Hawley's revision of the ecological approach, there would be no need to make any major modifications of the social disorganization approach to address issues of criminal opportunities and victimization since both perspectives share a focus on systemic control/guardianship. In fact, as we will see later in this chapter, the disorganization framework has been successfully used in such research (see Smith and Jarjoura 1988). However, Hawley's ecological framework makes an innovative contribution to the routine activities model through its implications concerning the convergence of offenders and non-offenders in space and time.

Just as Shaw and McKay presented evidence that the regulatory capacities of communities were differentially distributed in space, a large body of research indicates that targets of crime do not have a random distribution (see, for example, the pioneering work of Schmid 1960a, 1960b). However, the notion of space in the routine activities approach is much more subtle than the simple geography of targets. Rather, it refers to the concurrent location of targets and offenders. This orientation is drawn from Hawley's concept of the "friction of space" (1950:236–237), which refers to the time and energy required to transverse a line of travel in the course of sustenance activities. In some types of symbiotic relationships, technological developments have made physical distance relatively unimportant. For example, it is now easy to engage in almost instantaneous long-distance collaborative research through computer and facsimile machine telecommunications; in the very recent past, the best one could hope for was an overnight delivery of some desperately needed output.

However, since the routine activities model restricts its focus to those crimes which involve direct contact between the offender and the target, physical distance is a key factor that affects the likelihood that a criminal opportunity will actually be exploited. Holding all else equal, a convergence between motivated offender and suitable target is much more likely when the costs of travel are minimized. Therefore, as we will see in the next section, a neighborhood's proximity to a population of motivated offenders is assumed to be strongly related to the probability that the criminal opportunities of that community will be exploited by those offenders. Such an assumption receives strong support from the "travel to crime" literature, which has generally found that the probability of an offense declines with increasing distance from the residence of the offender (see Costanzo et al. 1986).

The second innovation that Cohen and Felson draw from the work of Hawley concerns the temporal aspect of crime; i.e., the offender and the victim have to be in the same location at the same time for an offense to occur. Hawley (1950:289; see Cohen and Felson 1979:590) argues that three temporal components are reflected in a community structure: the periodicity with which events occur (rhythm), the number of events that occur per unit of time (tempo), and the integration of the various rhythms within the community (timing). To the extent that the rhythm, tempo, and timing of events are predictable characteristics of a community, those events are considered to be routine activities, i.e., "habitual patterns of daily time use" (Rengert and Wasilchick 1985:33).

Rengert and Wasilchick (1985:Chapter 2) have provided perhaps the most extensive consideration of the role of time in criminal activities. A person's daily routine tends to be structured by nondiscretionary blocks of time. For example, assume that persons are required to be on the job from 8 to 12, and from 1 to 5. Unless they do not mind the possibility of dismissal from their jobs, those are nondiscretionary time blocks. Although the workers might like to see a movie during the lunch hour, they cannot. As Rengert and Wasilchick have documented, the key to a successful career in burglary is the ability to predict the structure of nondiscretionary time blocks during which a household is empty. Therefore, holding all else constant, one would expect the rates of criminal victimization to be highest in those neighborhoods in which a relatively significant portion of the population has predictable, nondiscretionary time blocks outside of the home, thereby creating an absence of capable guardians.[4]

It is important to realize that potential offenders also are characterized by a set of routine activities and often select their targets on the basis of information that they gather through recreational and work travel patterns.[5] For example, evidence exists that low-income individuals travel less extensively through urban areas than persons of other socioeconomic statuses (see, for example, Williams 1988). Therefore, it has been suggested that a great deal of spontaneous crime occurs in low-income areas because residents with high levels of motivation for illegal behavior note opportunities for crime within their own neighborhoods as they go about their routine activities that are relatively restricted to those areas (see Rengert 1989:166). However, since the routine activities approach emphasizes characteristics of the target rather than the offender, these dynamics are not addressed in the model (see the criticism of Jensen and Brownfield 1986).

Overall, despite the different emphases of Hawley and Park and Burgess, the primary ecological distinction between the routine activities and social disorganization models of crime reflects the integration of space/time considerations into the broader model. Nevertheless, the criminological implications of this relatively simple addition can significantly enhance our understanding of the relationship between neighborhood dynamics and criminal behavior.

The Basic Model

As noted in the preceding section, the routine activities model assumes that opportunities for criminal behavior go unrealized unless three elements converge in time and space: a motivated offender, a suitable target, and an absence of capable guardians. The elimination of any of these elements from a particular situation is sufficient to prevent the successful completion of a crime (Cohen and Felson 1979:589). For example, imagine a situation in which a neighborhood has a large number of unsupervised youths, all of whom are motivated to rob a convenience store. If that community has a strictly residential composition, then it is impossible to commit such robberies in the area owing to the lack of suitable targets.

The components of the routine activities model are not as clear-cut as they may appear for their relationships are often "hopelessly intertwined" in the real world (Hough and Lewis 1989:30). Felson and Cohen (1980:393) note that target suitability has at least four dimensions: value (either monetary or symbolic), visibility, access, and inertia (i.e., factors that are difficult to overcome for illegal purposes, such as bulky property). In later work (see Cohen et al. 1981), a distinction is made between attractiveness (the value and inertia dimensions of target suitability) and exposure (the visibility and access dimensions).

The concept of guardianship presents some special difficulties. In their 1979 and 1980 papers, Cohen and Felson seem to imply that guardianship is a primarily human activity, noting that it involves "police action . . . [and] . . . guardianship of ordinary citizens of one another and of property" (1979:590). Yet, in the 1981 paper of Cohen et al. (p. 508), this dimension has been broadened to include such objects as burglar alarms, locks, and barred windows. This creates a great deal of conceptual ambiguity in the distinctions between guardianship and attractiveness (since inertia appears to be a common element to each) and guardianship and exposure (due to the common element of access). In addition, guardianship and attractiveness components are completely confounded in personal crimes, for an attractive object that is being exposed (i.e., the person or property that may be forcibly taken from that person) is simultaneously acting as its own guardian. Those same characteristics that may make a person a relatively ineffective self-guardian (such as a small size or an advanced state of inebriation) are the ones that partly determine that person's status as an attractive target.

Cohen and Felson (1979:590; see also Felson and Cohen 1980:392) argue that the analytical distinction between target and guardian is not important in such cases. However, targets and guardians are not fully equivalent in the case of personal crimes. For example, neighbors routinely keep an eye on the activities of those who are very ineffective self-guardians, such as young children or the elderly. Therefore, the distinction is still central to the dynamics of criminal victimization. Unfortunately, the full implications of the target/guardian overlap in personal crimes has not received a great deal of attention (one small footnote in the Cohen and Felson, and Felson and Cohen papers), which has led to some confusion concerning the theoretical implications of many empirical findings.

Perhaps the most difficult logical aspects of the routine activities framework have involved the specification of the dynamics related to the convergence of the offender and target in space and time, especially since the focus of the model is restricted only to the characteristics and/or life-styles of the victim. Cohen and Felson (1979:591–593) suggest that the three most central concerns should be the timing of work, schooling, and leisure. Most subsequent work, therefore, has emphasized the amount of time that persons spend away from home in these activities, thereby concurrently decreasing the level of guardianship of household property and increasing the likelihood that a potential victim and offender might spatially converge "on the street." Therefore, although guardianship and exposure are conceptually distinct, it becomes very difficult to separate their effects.

The allocation of time to these various activities is unrelated to the risk of victimization if the probability of a direct contact with a motivated offender is zero. Although the offender receives little direct attention in the routine activities framework, Cohen, Felson, and their colleagues logically derive a set of predictions concerning the dynamics related to offender/target convergence. Their most important development in this respect is the introduction of the concept of proximity into the model (Cohen et al. 1981:507), i.e., the physical distance between areas where potential targets of crime reside and areas where relatively large populations of potential offenders are found. Since the probability of criminal behavior has been shown to decrease with distance from the offender's place of residence, those potential targets that are located a relatively short distance from areas in which a large number of motivated offenders reside are expected to be victimized at a higher rate.

As we will note later in this chapter, most of the research within the routine activities tradition has focused on the individual dynamics of victimization. As a result, very little attention has been paid to the urban dynamics that may affect the distribution of criminal opportunities among neighborhoods. However, Cohen and Felson (1979:593) clearly indicate that at the heart of the model is a focus on structural changes that affect the nature of sustenance-related activities, which in turn affect the distribution of opportunities. In particular, they argue that since World War II there has been an important shift in the location of routine activities from within the home to outside of it. A key indication of this process is the growth in the number of households that are empty for significant portions of the day owing to such societal changes as the increasing rate of labor force participation among women and the growth of single-adult housholds (see Cohen 1981:141). Residents of such households have a decreased ability to act as a guardian of their own property as well as the property of neighbors (see Felson 1986). This orientation has been most fully developed by Sampson (1986) and has resulted in a focus on the ability of particular family structures to act as agencies of social control and supervision (see the related discussion of his work in Chapter 2).

In sum, the basic routine activities model predicts that the rate at which criminal opportunities are exploited within a neighborhood is a function of the community distributions of exposure, proximity, guardianship, and target attractiveness, which in turn are functions of the societal structure of sustenance-gaining activities (see Figure 3–1). The intrinsically multiplicative nature of this model is worth highlighting. A criminal event is not a function of the presence of motivated offenders or the presence of suitable targets or a lack of capable guardians; rather, the word "and" must be substituted for the word "or" in the preceding function. This is not a simple academic exercise in semantics. A community may undergo a rapid and dramatic change in its population composition, which in turn leads to a decreased ability to exercise systemic social control. The social disorganization model would predict an increased probability that the residents of that neighborhood engage in criminal behavior. However, if the number of suitable targets for a particular form of crime in that area is zero, the routine activities model predicts that no criminal events of that type will occur within the community.

On the other hand, a community may be very stable, with a small population of criminally motivated residents and an effective system of systemic control. Yet, although the number of offenders and the level of control may remain constant, if the number of suitable targets in the community increases, the number of crimes committed in that area should increase (see the discussion of Cohen and Felson 1979:589). Thus, what appears to be a fairly simple and parsimonious model of criminal events leads to the prediction of a relatively complex series of interdependencies that lead to those events.

Figure 3–1. The basic routine activities model.

Collecting Indicators of the Routine Activity Model

As we noted earlier in this chapter, the routine activities model is assumed to represent a characterization of the group dynamics related to the distribution of criminal opportunities. Unfortunately, neighborhoods have not been utilized as the unit of analysis in such research nearly as often as one might assume, especially given the ecological nature of its theoretical roots. Rather, the preponderance of work in this area has either utilized the framework to make predictions concerning the likelihood of individual or household victimization (such as Miethe et al. 1987) or has focused on much larger aggregates than neighborhoods (see Cohen, 1981, Cantor and Land 1985 or Messner and Blau 1987).

Three general forms of data have been used to test the routine activities approach at the neighborhood level: officially collected records of the locations and nature of reported crimes[6] (see, for example, Block 1979 or Roncek 1981, 1987a), large-scale, national self-reported datasets that have been collected on a relatively ongoing basis, and locally collected, self-reported studies of victimization (Greenberg et al. 1982c; Taub et al. 1984; Skogan and Maxfield 1981). While all three sources of data have provided important insights into the dynamics of criminal opportunities, the national datasets have easily drawn the most attention among researchers. Therefore, it is useful to examine their utility for the study of neighborhood dynamics and criminal opportunities.

The National Crime Survey

The National Crime Survey (NCS) is an ongoing, annual national survey of approximately 100,000 households (representing around 200,000 individuals) currently sponsored by the Bureau of Justice Statistics that has been conducted since 1973 (for extensive reviews and discussions of the NCS, see O'Brien 1985 or Garafalo 1990). During the course of this survey, the respondents are asked a series of questions pertaining to the number and nature of victimizations that they have experienced during the preceding six months, and a limited number of questions concerning the location of the incident, the approximate time of day, and the presence of observers. The NCS is by far the most popular source of data that has been used to study victimization within the United States. Since the strengths and weaknesses of the NCS have been widely discussed elsewhere, our concern in this section is with the potential of these data to provide reliable information concerning the neighborhood dynamics of victimization.

Unfortunately, as Garafalo has discussed (1990:91–93), this potential is extremely limited. Given the fact that the households are drawn from a national sample, the expected number of survey respondents that would reside in a particular neighborhood of a particular city is essentially zero. Even if several respondents live in the same community, the small sample sizes would make the sampling errors astronomical. Thus, one could not place any confidence in any rates that had been estimated, especially since many of the criminal offenses are very rare events.[7] In fact, Garafalo argues that meaningful inferences cannot even be drawn at the level of individual states.[8]

Although it is impossible to create victimization rates for particular neighborhoods using the NCS, the dataset does contain information pertaining to fifty-five characteristics of the ennumeration district in which the sampled housing unit is located (see Garafalo, 1990:92). While such data do not enable an analysis of the neighborhood dynamics upon which we have focused in this book, it is possible to aggregate respondents who reside in similar neighborhood contexts, compute the mean victimization rates within those contexts, and conduct analyses on that basis (see, for example, Pope 1979 or Sampson and Castellano 1982).

Such uses of the NCS data have two serious drawbacks. First, the census data reflect characteristics of the area in which the sampled housing unit is located, and not of the neighborhood in which the victimization actually took place. Sampson and Castellano defend their use of these data in light of this limitation by arguing that most criminal events occur relatively nearby the residence of the victim. While this is true, the correspondence is far from exact as noted in the opening section of this chapter, and it is impossible to determine how many of the victimizations reflected in the NCS actually occurred in the neighborhood of the target.[9] While the NCS data have the advantage of reflecting criminal events that have not come to the attention of the police, we feel that the lack of data pertaining to the area in which target and offender converge makes the data far from ideal for the analysis of this issue.

In addition, the neighborhood data that have been integrated into the NCS are drawn from the 1970 U.S. Census. Thus, this information is significantly out of date, especially in those areas of the country that have experienced a great deal of economic and social change (see Garafalo 1990:93). Overall, therefore, the NCS data have provided only limited and potentially misleading insights into the neighborhood dynamics related to criminal opportunities and victimization.

However, there is one rarely analyzed component of the NCS data that may in fact provide relevant neighborhood-related information. For each criminal event that comes to the attention of the interviewer during the screening portion of the survey, the interviewer is required to write a brief, but detailed, narrative of the nature of the situation that led to the victimization. This information can provide a wealth of data not otherwise available in the NCS and has been used, for example, to examine the dynamics of school-related victimizations (see Garafalo et al. 1987). While it is possible that these narratives contain information concerning the relationship of neighborhood dynamics and victimization, the coding of such data is a relatively complicated procedure (see the discussion of Garafalo et al.). To the best of our knowledge, they have not yet been used in studies in which neighborhoods are the primary focus.

The Victim Risk Supplement

Given the ambitious nature of the NCS, it is to be expected that certain modifications in the original study design would eventually be necessary. In 1978, the Bureau of Justice Statistics contracted with the Bureau of Social Science Research, Inc., to form a consortium that would develop, test, and negotiate improvements to the NCS (see Lynch, 1990). One of the most important of these was the creation of the Victim Risk Supplement (VRS), which was administered during the February 1984 wave of the NCS. The primary goal of this supplement was the collection of additional data concerning crime prevention measures that were taken at home and at the workplace, and about perceptions of the safety of homes, neighborhoods, and places of work (Bureau of Justice Statistics, 1986). The VRS collected data from approximately 11,000 households, representing over 21,000 individuals. A special attempt was made to obtain information concerning the dynamics of victimization within the workplace.

The availability of the VRS data represented a significant improvement in the ability to analyze situational factors that increase the likelihood of offender-target convergence (see Lynch 1987; Maxfield 1987). Yet, like the NCS, the data are most suitable for the analysis of individual-, as opposed to community-level, dynamics. However, after leaving each household, the interviewers were required to record information concerning neighborhood characteristics that may influence the risk of victimization. Therefore, it is possible that future uses of the VRS may provide relevant insights to the group dynamics we have discussed in this chapter.

The British Crime Survey

From the standpoint of neighborhood researchers, it is regrettable that two of the major data collection efforts of the U.S. Department of Justice can provide relatively little information concerning the local community distributions of criminal opportunities. However, this is not the case with the British Crime Survey (BCS), which has been receiving an increasing degree of attention from neighborhood researchers in the United States (see Sampson and Wooldredge 1987; Sampson and Groves 1989; Sampson and Lauritsen 1990).

The BCS is a national victim survey sponsored by the Home Office

Research and Planning Unit that was conducted in 1982 and 1984 (see Hough and Mayhew 1983; S. Smith 1986:Chapter 3). This dataset has three major advantages over the NCS and the VRS. First, an enormous amount of information is available concerning the life-styles of the respondents, including work and leisure-related routine activities, the temporal structure of household occupancy, property contained in the household, and modes of transportation (see Hough 1987). Second, the BCS also contains data concerning the criminal activities of the respondents themselves, thereby enabling a researcher to determine the dynamics of the relationship between offender and victim statuses.

Most important, the BCS contains a wealth of information concerning the neighborhood context of the respondents. For example, the social disorganization research of Sampson and Groves that was discussed in Chapter 2 was able to derive a number of indicators of systemic control on the basis of the material contained in the BCS. Such data, which are key indicators of the guardianship component of the routine activities model, simply are not available in either the NCS or the VRS. In addition, each of the individual records contains a code pertaining to the residential electoral wards of the respondents. These wards are relatively compact and average about 5,000 residents; 238 wards are reflected in the BCS, with an average of 46 respondents per ward (Sampson and Wooldredge 1987:375). Not only does this enable a researcher to integrate current neighborhood-related census material into the dataset, but it also becomes possible to use the ward identification to aggregate individual responses to the survey into meaningful community measures.

While the capabilities of the BCS are extremely attractive, the data must be used with some caution. As with the NCS, the number of respondents who have been the victims of particular types of crime is quite small. Therefore, the crime-specific sampling errors tend to be large, which makes it difficult to compare differences between neighborhoods (see the discussion of S. Smith 1986:54). As a result, most researchers, such as Sampson, Wooldredge, and Groves, tend to aggregate rare individual offenses into broad categories so that the rates are more reliable.

A second limitation is that all the respondents interviewed in the BCS are aged sixteen or over, compared to the twelve or over population represented in the NCS. Therefore, the relatively large number of victimizations experienced by people aged twelve to fifteen are not reflected in the data. Finally, the respondents in the BCS are asked to report on events that occurred during the preceding year (approximately), as opposed to the six-month time period of the NCS. Research by Turner (1972) has shown that the probability that a victimization is reported to the interviewer declines with the time from the event. For example, he reports that only 30 percent of those events occurring ten to twelve months prior to the interview are actually reported. This characteristic of the BCS also leads to an under-counting of criminal events. Nevertheless, the BCS provides the best available national data amenable to the examination of neighborhood dynamics and criminal opportunities.

Operationalizing the Model

Despite the relative theoretical simplicity of the routine activities model, it has been notoriously difficult to collect reliable and valid indicators of its central components. In fact, even the creation of an appropriate crime rate has presented special difficulties. For example, since the focus of the social disorganization model is on the rate of offending among the residents of a neighborhood, the computation of the crime rate involves a simple ratio of the number of offenses committed by the residents of a neighborhood divided by the number of residents. Most studies in the routine activities tradition have operationalized victimization rates similarly, i.e., the number of criminal events that occur in an area divided by the number of residents at risk. However, the routine activities model is concerned with the exploitation of potential opportunities for criminal behavior. Therefore, the number of criminal events must be standardized by the number of targets that present an opportunity for crime in an area, not by the size of the residential population (see Clarke 1984).

An example may help clarify this important distinction. Let's assume that we are studying two neighborhoods, each with a population of 100 people. For whatever reason, the residents of neighborhood A own a total of 150 cars, while the residents of area B own only 10; during the period under examination in our research, five cars are stolen in each community. Obviously, criminal opportunities are being exploited at a much higher rate in area B. However, a population-based measure would lead to the erroneous conclusion that the rates of victimization were equal. Keith Harries (1981:151) cites a 1976 study of Wesley Skogan that documents just such a situation: while the vehicle theft rate per 1,000 residents was higher in Los Angeles than New York, the rate per 1,000 vehicles was lower.

The situation is even more complicated than suggested by this simple example, for the number of opportunities for crime provided by a given target in any specified period of time is not constant (see Clarke 1984:76). Perhaps most of the residents of neighborhood A drive to work, while most of the residents of neighborhood B take the bus. In such a situation, ignoring any uses of the automobile for leisure activities outside of the community, the number of opportunities for victimizing any particular car in area B is much greater than in area A. If it were possible to create an index of the amount of time that targets presented opportunities for criminal behavior (rather than focusing on the number of targets per se), we would have to make further modifications in our conclusions concerning the differences in the rate of criminal events in the two areas.

There have been many attempts to address this significant complication through the computation of "risk-related" rates in which the denominator represents the number of opportunities for crime (see the review of Harries 1981). One of the most influential efforts was that of Sarah Boggs (1965:900–901) who used, for example, the ratio of business to residential land use in a neighborhood as the denominator for the computation of the business robbery rate. While the correlations of her risk-related rates with the traditional population-based rates were very high for crimes against the person, this was not the case for property offenses. In fact, the correlations between the two sets of rates were actually negative for nonresidential night burglary, grand larceny, and nonresidential day burglary. Therefore, the use of an inappropriate crime rate can lead to misleading conclusions concerning the relationship between neighborhood dynamics and criminal opportunities.

It may appear that the population of a neighborhood would be the appropriate denominator for risk measures pertaining to crimes against the person. While the use of the traditional rate is certainly more justifiable in the analysis of such criminal activities, it is important to reemphasize that the routine activities model focuses on all the criminal opportunities afforded by a neighborhood, regardless of their source. Certain areas of a city are characterized by a relatively high degree of nonresidential use, such as the central business district or a neighborhood that provides many opportunities for the use of leisure time. At the same time, some residents of those areas are involved in activities (such as work or school) that take them out of the neighborhood for significant parts of the day, thereby eliminating them from the population at risk during that time.

The complicated issue of how and where people allocate their time has received considerable attention in the routine activities literature. For example, Cohen and Felson (1979:594–595) provide estimates of the victimization per billion person-hours spent in three general types of location (home, street, and elsewhere). More recently, Lynch (1987) utilized the VRS to specify four primary domains of routine activities: work, school, home, and leisure. Likewise, the information contained in the BCS enabled Sampson and Wooldredge (1987) to examine the number of hours per week that a respondent's house was empty during the day, during the evening, and due to trips outside of the county. However, such refinements primarily have been used to create indicators of guardianship and target attractiveness; the more basic measurement issues concerning the creation of the victimization rates themselves have not been addressed. In fact, nearly all the neighborhood-level studies of victimization that will be discussed later in this chapter utilize crime rates that are standardized by population rather than by opportunities. Such a limitation has been a major hindrance to the theoretical refinement of the routine activity approach.

On the other hand, a great deal of attention has been paid to the measurement of the four primary components of the routine activities model. In fact, Maxfield (1987a:278) notes that one of the most common themes in this body of literature is the inadequate measurement of the central variables (see also Massey et al. 1989). As noted above, there are logical ambiguities in the conceptualization of the components of the framework; these are often reflected in the indicators that have been used to operationalize the model. For example, the "life-style" variable utilized in the research of Cohen et al. (1981) combines measures of household composition and labor force participation, thereby confounding exposure to crime and guardianship (McCraw 1986).

However, the selection of appropriate and conceptually distinct indicators is not simple, for the information needed to rigorously measure the routine activities components is not available in many datasets. In this respect, the limitations of the model are very similar to those found in the social disorganization research that has tried to operationalize systemic control strictly on the basis of census data. For example, the percent of the population aged fifteen to twenty-four, the percent of households classified as primary individual, and the weight in pounds of the lightest television set advisertised in the Sears catalog represent all the explanatory variables incorporated into Felson and Cohen's (1980:400) routine activities model of burglary rates in the United States.

The most important advances in the measurement of the routine activities components have occurred in studies that have restricted their focus to a small number of neighborhoods. Three of these studies in particular have collected especially rich bodies of data. The Reactions to Crime Project, directed by the Center for Urban Affairs at Northwestern University between 1975 and 1980, surveyed the residents of ten Chicago, Philadelphia, and San Francisco neighborhoods (see Skogan and Maxfield 1981; Lewis and Salem 1986; Skogan 1990); the concepts of neighborhood disorder and incivilities that were discussed in Chapter 2 have been developed extensively within this body of research. Although this work has primarily focused on the link between victimization and the fear of crime (which will be discussed in the next chapter), the project also collected information relevant to the examination of guardianship, as reflected in indicators of neighborhood integration, processes of informal social control, and the extent of social and physical disorder within the neighborhoods.

Excellent indicators of exposure, guardianship, target attractiveness, and the spatial use of time are available in the Chicago Neighborhood Study, directed by Richard Taub and his colleagues (see Taub et al. 1984). During 1979, 3,310 residents of eight Chicago neighborhoods were interviewed; the neighborhoods were chosen to allow comparisons between high- and low-crime areas, areas with changing and stable racial compositions, and areas with appreciating and stagnating housing markets (Taub et al.:18–19). In addition to questions pertaining to victimization and the fear of crime, the Taub study solicited information pertaining to the nature, timing, and location of routine activities, satisfaction with and attachment to the neighborhood, perceived problems of the neighborhood, steps taken to address those problems, the extent of formal and informal relational networks within the community, the perception of signs of disorder, sources of crime-related information, and the level of involvement in crime prevention activities. These data were supplemented with block-level police reports, census data, and independent ratings of the quality of the neighborhood commercial areas, and the general appearance of the homes and neighborhood in general.

Finally, the Atlanta Community Study (see Greenberg et al. 1982b, 1982c) conducted interviews with 523 residents of three pairs of adjacent neighborhoods in Atlanta that were selected for comparable racial composition and comparable economic status; one member of the pair had a high rate of reported crime while the other had a low rate. The primary focus of the Atlanta study was on "territoriality," i.e., the maintenance of the social control of crime within a geographically specified area, a concept with obvious similarities to guardianship and systemic control. A wide variety of information was collected from the respondents pertaining to territoriality, such as identification with the neighborhood, the degree of neighboring, the network of local social ties, emotional attachment to the area, the use of neighborhood space, the degree of perceived control over events in the community, the ability to distinguish between strangers and residents, patterns of informal surveillance, and the probability of intervention if a crime was witnessed. These data are supplemented with information collected by the Atlanta Planning Commission (the PLAN file) pertaining to the physical characteristics of every parcel of land in the study site.

Perhaps due to its general (and unwarranted) reputation as a model of individual and household victimization, neighborhood-level analyses of victimization rates tend to utilize indicators that are only partly relevant to the routine activities framework. Rather, as we will see in the next section, many of these studies are dominated by variables intrinsic to the social disorganization perspective. Since many offenses are committed in the residential neighborhood of the offender, it is only logical to assume that the same processes of disorganization that are related to the distribution of offenders in a community will be related to the exploitation of criminal opportunities in an area. And since the concepts of systemic control and guardianship are essentially identical, tests of a disorganization model of victimization are also partial tests of the routine activities model. Yet at best this provides an incomplete explanation of victimization, for the disorganization perspective is not suited to address the central dynamics through which offenders and targets converge in those neighborhoods. However, there have been some attempts to at least partially integrate these complementary approaches into a single model.

The Relationship between Neighborhood
Characteristics and Criminal Events

Since victimization surveys are a relatively recent development within criminology, the classic studies of criminal opportunities relied on official records of offenses that were reported to the police or, occasionally, arrests that were made.[10] One of the earliest and most ambitious of these studies was Calvin Schmid's (1960a, 1960b) examination of the neighborhood distributions of offenses in Seattle between 1949 and 1951. In a key section of the analysis (1960b:670–674), Schmid classifies each of Seattle's local communities in terms of the three dimensions of community structure identified by Shevky and his colleagues (Shevky and Bell:1955): urbanization, segregation, and social rank. The segregation (in the racial and ethnic composition of the area) and social rank (educational and occupation composition) variables reflect community characteristics traditionally analyzed in social disorganization research. However, the three components of the urbanism dimension have been widely used as indicators of the structure of household activities: the fertility ratio, the proportion of single-family dwelling units, and the proportion of women in the labor force. Schmid finds (1960b:673) that this dimension is related to neighborhood rates of all forms of larceny, fraud, burglary, and robbery with the exceptions of bicycle theft and nonresidential burglaries.

There are some important limitations of the Schmid research. For example, the neighborhood characteristics considered in his analysis are limited to those available from census materials. Therefore, despite the large number of variables that are incorporated into his research (see 1960a:534), there are no direct indicators of any of the four routine activity components with the possible exception of a single variable pertaining to the percentage of dwelling units in a neighborhood with television sets (which could reflect target attractiveness). Likewise, although his operationalization of household activities is similar to that found in some more recent work, it provides a very indirect measure of the time spent away from the household. In this respect, his work is similar to census-based social disorganization research in which indicators of systemic control have not generally been available to analysts. In addition, the rates used by Schmid are all population-based and, therefore, are not standardized for the number of criminal opportunities in the neighborhood. Nevertheless, this research provided a significant empirical context for the later development of the routine activities approach.

As noted earlier in this chapter, the research of Sarah Boggs (1965) introduced an important extension of the Schmid research through the computation of crime rates that were standardized by the opportunity structure of the community. However, like Schmid, her analysis did not consider any variables that could be considered direct indicators of the routine activities component. Boggs also utilizes the Shevky-Bell typology to characterize the census tracts of St. Louis during 1960. While the urbanization dimension is a significant predictor of neighborhood rates of auto theft and robbery (with the exception of the robbery of businesses), it is not related to the other crimes examined in her analysis (burglary, homicide-assault, and rape).

More recent research based on reported crime tends to support the findings of Schmid and Boggs concerning the effects of the urbanization factor. For example, a series of studies of Cleveland and San Diego by Dennis Roncek (1981 1987a, 1987b) indicates that the percent of the households on a city block that are occupied by a single individual or headed by a female is related to the number of violent and property crimes that occur in that area. In addition, Richard Block (1979) has shown that the ratio of families earning more than three times the poverty level to those earning 75 percent or less (which some have interpreted as a rough indicator of the routine activity proximity component) was significantly related to population-based rates of homicide, robbery, and aggravated assault in Chicago's neighborhoods during the mid-1970s.

Overall, while research based on official crime reports is generally consistent with the predictions of the routine activities model, the reliance on census data for the creation of neighborhood variables results in a limited ability to examine the dynamics through which criminal opportunities emerge within a local community. However, these studies do suggest that the distribution of such opportunities reflects a complex mixture of the dynamics of both the social disorganization and routine activity models.

Owing to an increasing dissatisfaction with the scope of the neighborhood information available from the census, survey data have become the most popular source of information analyzed in studies of criminal opportunities, especially after the initiation of the National Crime Survey. We noted earlier in this chapter that there are some severe limitations in the inferences that can be made at the neighborhood level using the NCS. Nevertheless, several studies have examined the neighborhood context of victimization on the basis of these data.

The earliest of these studies (such as Pope 1979; Sampson and Castellano 1982; Sampson 1983) restricted their focus to a relatively small number of variables relevant to both the routine activities and social disorganization approaches. Pope, for example, presents evidence based on the 1973–1976 NCS that a high percentage of female-headed households in a neighborhood is positively related to high rates of auto theft, larceny, and (especially) burglary victimizations. Likewise, the research of Sampson and Castellano documents significant relationships between two indicators of economic composition (the proportion of family incomes less than $5,000 and the unemployment rate) and rates of theft and violence victimizations, while Sampson finds a similar relationship involving the percentage of dwelling units in a neighborhood that are located in structures containing five or more units.

Later NCS research of Sampson (1985, 1986) has utilized a much broader set of neighborhood-based variables. In his 1986 paper, for example, he examines the relationship between the rates of theft and violent victimizations in a neighborhood and the unemployment rate, income inequality, the racial composition, the level of residential mobility, structural density, and three indicators of family structure: the percent divorced or separated, the percent of families that are female-headed, and the percent of primary individual (single person) households. He presents evidence supportive of both the routine activity and social disorganization perspectives, including the significant finding that the racial composition of a neighborhood is not related to the rate of violent victimization once the effects of the female-headed household variable are controlled.

Unfortunately, the structure of the NCS limited the types of analysis that Sampson could attempt. To create the neighborhood variables, he trichotomized each of the relevant indicators and computed three sets of victimization rates within each category: one for 1973, 1974, and 1975. This small number of observations restricted him to the simultaneous consideration of at most three variables at a time. Perhaps owing to a general frustration with the analytical problems that arose due to the structure of the NCS and the inferential problems that have been noted, the 1986 Sampson paper is the last major NCS-based neighborhood level study of victimization of which we are aware.

There is certainly a third reason for the limited neighborhood research that has been conducted with the NCS since the mid-1980s: the availability of the British Crime Survey. Some of the findings that have been based on the BCS confirm the patterns suggested by the NCS. For example, the rate of burglary victimization is higher in neighborhoods characterized by a high percentage of primary individual households (Sampson 1987b).

However, while the BCS dataset contains much more detailed and reliable neighborhood information than the NCS, most research using those data continue to be characterized by a limited operationalization of the dynamics central to the routine activities model. For example, in Chapter 2 we discussed the findings of Sampson and Groves (1989) concerning neighborhood dynamics and the likelihood of criminal involvement on the part of the residents. Sampson and Groves also extensively examine the relationship of social disorganization and victimization. In general, they show (pp. 787–789) that high levels of disorganization are related to high neighborhood rates of victimization. However, they do not directly address the routine activities model in this paper, although they present evidence that a scale of family disruption (which combines the proportion of divorced or separated adults and the percentages of single adult households with children) is related to all forms of victimization except vandalism (see also Sampson 1987b).

Similar findings using data drawn from fifty-seven neighborhoods in Rochester, New York, Tampa–St. Petersburg, Florida, and St. Louis have been published by Smith and Jarjoura (1988). After controlling for the effects of the social disorganization variables included in their models, they find that the percentage of single-parent households, the percentage aged twelve to twenty, and the population density are strongly related to the rate of burglary victimizations in these communities. The single-parent household indicator, however, was not related to the rate of violent criminal events. In addition, they replicate the finding of Sampson (1986) that the racial composition of a neighborhood is not related to rates of victimization after controls are made for the social disorganization and routine activity components of the model.

Although the studies that have been discussed are all characterized by a limited operationalization of the routine activities model, the findings have been fairly consistent with its predictions concerning guardianship. However, Smith and Jarjoura (1989) provide an important caution concerning the conclusions that may be drawn from such analyses. Just as was the case with the majority of social disorganization research, the use of compositional variables in routine activities models confounds individual and group-level effects. Therefore, what we may be observing is the neighborhood concentration of individuals who have life-styles conducive to victimization, and not the effects of any form of group dynamics per se.

Two recent papers [11] have examined this proposition through the use of contextual analysis and both have generated results that confirm the existence of relevant neighborhood-level dynamics (see the illustrative results of Smith and Jarjoura 1989 in Table 3–1). An important feature of the Smith and Jarjoura study design is its focus on household burglaries, which automatically restricts the criminal events to those which occur within the residential neighborhood. This research clearly illustrates the complexity of the victimization process, for as shown in Table 3–1, while high-income neighborhoods are less likely to be victimized, high-income households are more likely to be selected for victimization within a neighborhood.

The most complete routine activities analysis of the individual and neighborhood dynamics related to victimization is found in Sampson and Wooldredge (1987), who utilize the indicators collected by the BCS to develop measures of target attractiveness, guardianship, and exposure at both the individual and community level. As illustrated in Table 3–2, they provide convincing evidence that even when the characteristics of individual life-styles are controlled, the community context of criminal opportunities has a significant effect on the likelihood of experiencing a household burglary. Although there are some differences in the patterns characterizing the four types of victimization examined in their analysis (burglary, household

Table 3-1
Contextual Model of Houshold Burglary Victimization

	M.L.E.[a]	t
Houshold characteristics		
Couple	0.041	0.31
Single parent	0.737	5.58
Single male	0.356	1.74
Single female	0.303	1.61
Two males	0.768	2.44
Two females	0.109	0.33
Apartment	−0.131	−1.19
Household income	0.063	2.17
Nonwhite	−0.145	−1.23
Resident less than 3 years	0.004	0.04
Number of residents	0.107	3.34
Age of oldest household member	0.018	−6.00
Neighborhood characteristics		
Racial heterogeneity	1.116	3.54
Percent nonwhite	−0.005	−1.93
Population density	0.134	1.17
Residential instability	0.019	2.71
Percent of multiple dwelling units	0.002	0.54
Percent aged 12–20	0.041	2.93
Percent of single-parent households	0.048	2.11
Median neighborhood income (logarithm)	−0.477	−2.64
Social integration	−0.524	−2.50
Percent single-person households	−0.007	−0.32
Constant	1.568	
Likelihood ratio test	312.34	

Adapted from Social Forces V68(2), 1989 (p. 633). "Household Characteristics, Neighborhood Composition, and Victimization Risk by Douglas A. Smith and G. Roger Jarjoura, Copyright © the University of North Carolina Press.
[a]Maximum likelihood logit estimate.

theft, personal theft with and without contact, and personal theft within a fifteen-minute radius of home), the dominant family structure in the neighborhood has a guardianship effect in all the analyses. Likewise, as would be predicted by the routine activities approach, high levels of target attractiveness (as indicated by the percentage of households with VCRs) are related to burglary and individuals are most at risk of personal theft in areas with high levels of street activity.

Table 3–2

Risk of Victimization by Individual Characteristics and Community Context

	M.L.E.
Individual characteristics	
Age of head	−0.007[b]
Education	−0.142
One-person household	0.323[b]
Household has VCR	−0.004
Time household is empty	0.036[a]
Household appliances	0.088
Neighborhood characteristics	
Percent family disruption	0.059[b]
Percent primary individuals	0.016[b]
Social cohesion	−0.043[a]
Percent households with VCRs	0.012[a]
Percent households empty during day	−0.005
Percent unemployed	0.048[b]
Percent apartment households	0.012[b]
Model chi-squared	128.19 (13 df)

Adapted from "Linking the Micro- & Macro-Level Dimensions of Lifestyle-Routine Activity-Opportunity Models of Predatory Victimization" by Robert J. Sampson and John D. Wooldredge, in the *Journal of Quantitative Criminology*, Vol. 3 (p. 382), 1987 by permission of the author and Plenum Publishing Corporation.
[a]1.5 times S.E.
[b]2.0 times S.E.

The Sampson and Wooldredge study provides an extensive use of the neighborhood variables contained in the BCS. Nevertheless, the measures of the routine activities components are still relatively crude. This restriction in the number of relevant and reliable indicators is one of the drawbacks of conducting secondary analyses of large datasets like the BCS, which may not have been designed with neighborhood effects as a primary consideration.[12] In addition, although rich data pertaining to the neighborhood context of criminal opportunities may have been collected in a particular study, the dynamics of routine activities or social disorganization that lead to victimization may not be the primary substantive interest of the investigators.

For example, while the design of the Reactions to Crime Project discussed earlier in this chapter was clearly influenced by the social disorganization framework, there has been little consideration of the dynamics that may lead to an offender/victim convergence within the study of neighborhoods (see Skogan and Maxfield 1981; Lewis and Salem 1986; Skogan 1990). However, that such processes may be relevant to a full understanding of the levels of victimization experienced by these communities is suggested by the case of Visitacion Valley in San Francisco (Skogan and Maxfield 1981:105). Although this community was a relatively affluent, home-owning residential area, it was characterized by a high rate of victimization. Skogan and Maxfield note that this may be due to the presence of a nearby public housing project, which would support the predictions of the routine activities model concerning proximity.

Likewise, the primary use of the Chicago Neighborhood Study data has been to examine the dynamics through which fear and victimization are related to ongoing patterns of community decline (Taub et al. 1984). For example, Taub et al. (pp. 136–137) present evidence that high levels of vandalism have a negative effect on the perceived economic future of the area and, in turn, the financial investments that residents are willing to make in the local community. Thus, their work shows that the exploitation of criminal opportunities is not simply the outcome of the ecological processes that shape the characteristic routine activities of an area. Rather, they may also shape those processes.

The work of Taub et al. concerning neighborhood investment has led to significant advances in our understanding of the relationship between criminal victimization and neighborhood dynamics. Unfortunately, these data have not been widely used to study the convergence of offenders, suitable targets, and capable guardians despite the large amount of valuable information that is available. An important exception is found in the study of D. Garth Taylor et al. (1986), which finds that after controls are made for the level of victimization in the residential census tract and other neighborhood characteristics, personal actions that might limit the opportunities for potential victimization (such as avoiding public transportation, walking in groups, and installing security devices) have a very small effect on the likelihood of individual victimization. In addition, the neighborhood effect on individual victimization is not constant across all eight neighborhoods (see Taylor et al.:Table 4). Such initial findings suggest that the Chicago Neighborhood Study dataset has a great deal of untapped potential for understanding the neighborhood dynamics related to the exploitation of criminal opportunities.

On the other hand, Stephanie Greenberg and her colleagues (1982b, 1982c) have used the Atlanta Community Study to made extensive examinations of the neighborhood sources of criminal opportunities. Their findings have not provided especially strong support for a systemic control/ guardianship approach to the control of criminal opportunities. For example, while the low-crime study areas were more residentially stable than high-crime areas, this relationship was significantly attenuated when controls were made for the age structure of the neighborhood. In addition, they found no significant differences between high- and low-crime areas in the extent of neighboring, the scale of local friendships networks, the perceived similarity with neighbors, and the exchange of information among residents. However, the physical characteristics of the areas, such as the number of housing units per structure, the commercial use of land, street type, and physical insulation from surrounding areas, all significantly differentiated between high- and low-crime areas.

On the basis of these findings as well as those produced by related studies, Greenberg et al. (1985:I-100) concludes that informal control has not been shown to have an effect on the rate of reported crime within a neighborhood "with any reasonable level of confidence." Rather, they argue that the physical characteristics of neighborhoods have a direct effect on crime by affecting the access to outsiders, the degree of surveillability, the ease of entrance and exit, and the number of potential targets, offenders, and witnesses (i.e., guardians; p. I-101). Such findings support the predictions of the routine activities model concerning attractiveness, exposure, and proximity, but suggest a much more limited form of guardianship than that implicit to a systemic model of control.

The Greenberg et al. findings can be criticized on the basis of the limited number of neighborhoods, the failure to separate individual and contextual effects and the relative simplicity of the analysis, which tends to focus on univariate tests of significance (see, for example, Greenberg et al. 1982c). Most important, the sample of neighborhoods was intentionally designed so that the pairs of neighborhoods were geographically, racially, and socioeconomically similar. Given the ecological argument of the social disorganization model, it should be expected that these communities would have the similar levels of systemic control noted by Greenberg et al. Therefore, since these effects are essentially partialed out by the research design, it is not surprising that the neighborhoods in this study that are most commonly victimized are those which, holding the level of systemic control constant, provide the greatest ease of movement into and throughout the community.

This proposition receives support from the results of the Baltimore neighborhood studies conducted by Ralph Taylor and his colleagues (1985, 1986). Like Greenberg et al., they find strong neighborhood-level correlations between the physical characteristics and land uses of a neighborhood and the rates of crime in those areas. However, these two dimensions are also strongly correlated with the sociodemographic composition of the communities; when the effects of this composition on crime are controlled, the magnitude of the relationships described by Greenberg et al. are greatly reduced.

Conclusions

As we noted above, certain features of the routine activities model are very similar to those found in the social disorganization framework, especially those pertaining to guardianship and systemic social control. Therefore, it is not surprising that the theory of disorganization has been used in the study of community rates of victimization. However, the routine activities model goes beyond such an approach in its emphasis on the convergence of offenders and suitable targets. Unfortunately, it has been very difficult to operationalize these dynamics, forcing most neighborhood-level studies to utilize indicators that only partly specify the theory and whose reliability and validity are questionable. As a result, it cannot be said that any truly definitive test of the model has ever been made, and conclusions that have been drawn on the basis of the many studies discussed in this chapter must be considered with a great deal of caution.

It might be countered that although partial tests of the theory are not completely satisfactory, they do provide important steppingstones for the refinement of the model. In a sense this is true. However, the failure to consider the full set of dynamics embedded into the theory can lead to the fairly trivial conclusion that crimes occur in neighborhoods where there are opportunities for such crimes without any sense of the dynamics that make such convergences likely. Hopefully, future research will make use of the measurement innovations contained in such datasets as the Chicago Neighborhood Study to more fully explore the dynamic implications of this perspective.

4

Neighborhood Dynamics and the Fear of Crime

Like a nun, I kept custody of my eyes. Walking home at sunset a few days ago, I kept them averted from the two men who were urinating against the wall of the Spanish restaurant at the corner. I kept them averted from the hooker at the entrance to that Cloaca Maxima that is the meat market these days. I kept them averted from the stranger who was walking toward me from the river front. Perhaps he was simply on his way to the supermarket for a quart of milk. But, hey, you can't trust anyone anymore
—(Cantwell, 1990).

Throughout history, residents of urban areas have expressed fears about many conditions of their everyday lives (see Tuan 1979, for an historical perspective on urban fear). While it is generally assumed that one of the hallmarks of modern urban life is the fear of criminal victimization, the number of urban respondents who report that they are "bothered" by crime in their neighborhood is actually less than the number who note problems with traffic or noise (Bureau of the Census 1989:49).[1] Nevertheless, 40 percent of those interviewed in the 1989 General Social Survey indicated that they were afraid to walk alone at night in some areas within a mile of their homes (Flanagan and Maguire 1990:153), and political rhetoric that exploits such fears (as reflected in the notorious Willie Horton commercial used by George Bush in the 1988 presidential campaign) strikes a respondent chord among many citizens. Therefore, the emotions reported by Cantwell, which she recorded shortly after an apparently random shooting occurred in her middle-class neighborhood, are familiar to many residents of large cities.

Anyone who examines the fear of crime literature will be struck by several intriguing paradoxes. For example, it might understandably be assumed that the relatively high levels of reported fear that have characterized the United States in recent times reflect a perceived vulnerability to what former Attorney General Richard Thornburgh has described as "the carnage in our mean streets" (quoted in Krisberg 1991:141). Yet during 1989, only 29 violent victimizations occurred for every 1,000 persons over the age of

12 in the United States (Bureau of Justice Statistics, 1991). While this rate of crime is certainly a justifiable source of concern, it is far less than one would expect on the basis of Thornburgh's comments. Therefore, one paradox that has often been noted is that many more people are afraid of crime than one would expect given the risks of victimization. Likewise, a great deal of research has documented the finding that those least likely to be victimized (such as women and the elderly) report the highest levels of fear, while those populations most at risk (such as young, black men) report the lowest levels. As we will see, such findings indicate that the fear of crime is the outcome of much more subtle processes than a simple response to perceived risk.

The history of criminological inquiry into the fear of crime has its own interesting paradox. We noted in the last chapter that although the routine activities model is grounded in an ecological theory of group behavior, the implicit community dynamics have been relatively undeveloped in the empirical literature. Therefore, since fear is one of the most deep-seated, personal emotions experienced by humans, one would expect that such group dynamics would not receive a great deal of attention. However, the neighborhood context of fear has been one of the most important themes of this research tradition since the late 1960s, when the fear of crime emerged as a central consideration of criminology (see Biderman et al. 1967; Ennis 1967; Reiss 1967). To understand this development, it is necessary to examine briefly the concept of fear itself.

The Emotion of Fear

Although fear is one of the most basic of human emotions, it has been very difficult to define precisely. For example, although words such as fear and anxiety are often used synonymously, some psychologists differentiate between the two terms.[2] Likewise, some criminologists (such as Garafalo 1981 or Taylor and Hale 1986) have argued that there is a difference between the fear of crime and worry about crime.[3] In fact, Fred DuBow and his associates (1979:1) have noted that a wide range of different emotions and impressions have been subsumed in the literature under the common term "fear," such as perceived risk, concern, worry, and anxiety. Therefore, contradictory findings concerning the fear of crime can often be explained simply by the different ways in which fear has been conceptualized (DuBow et al. 1979:2).

DuBow et al. (pp. 2–6; see also Merry 1981 and Ferraro and LaGrange 1987) have identified three analytically distinct (but related) dimensions of crime perceptions that commonly have been used interchangeably in the literature as defining characteristics of fear. Some studies have asked questions pertaining to the seriousness of crime and the priorities that the political process should give to crime control. DuBow et al. argue that while these expressed values and concerns may be related to the fear of crime, they are not direct indicators of the emotional response that is assumed to be a defining characteristic of fear. Other studies have used perceived personal risks of victimization as a surrogate measure of fear. Again, while such judgments may be related to the fear of crime, they reflect a cognitive evaluation of environmental cues that may lead to the emotional reactions implicit to the concept (see Merry 1981:10).

The third general approach focused on the emotional response of fear itself, and this is the orientation we have taken in this chapter. Yet even within such a restricted focus, the analysis of fear within a neighborhood context presents special difficulties. As James Garafalo (1981:841) has observed, there is an important difference between actual and anticipated fear of crime. Chronic fear is experienced by a relatively small number of people; for most of us, it is triggered by some immediate cue. Therefore, Garafalo argues that a full understanding of the fear of crime is impossible without a determination of the types of situations and cues that have elicited such responses, how strongly people have experienced such emotional reactions in those situations, and how often they find themselves in such situations.

One of the central situational contingencies that is related to the likelihood of a fearful response is that of "familiarity" with an environment since it imparts at least a partial sense of control over potentially threatening situations (Rachman 1990). Since the neighborhood represents one of the most intimately familiar contexts that provides a basis for distinguishing between "us" and "people different from us," the degree to which the fear of crime represents the perception that a community has been invaded by "unknown strangers" (Ennis 1967:80) is one of the dominent themes of contemporary research. Thus, the fear of crime may reflect in large part a fear of people whose actions may be unpredictable or of stangers over whom one has little or no control (DuBow et al. 1979:4). In fact, some of the earliest major studies of victimization (such as Biderman et al. 1967; Ennis 1967) concluded that the fear of crime was primarily a fear of strangers.

Dan Lewis and Greta Salem (1986:8) have noted that this inference was primarily an ex post facto attempt to explain some paradoxes that were observed in the presumed relationship between victimization and the fear of crime. However, Sally Merry's (1981) study of a public housing project indicates that the role of the stranger is important. For example, she presents evidence that the residents of a housing project who are able to personalize neighbors involved in illegal behavior are less afraid of crime than those residents for whom criminal victimizations represent random attacks of anonymous origin. Obviously, the systemic structure of relationships in a neighborhood is a key determinant of the degree to which information circulates concerning the level, nature, and sources of local criminal activity.

The fear of crime also may be a highly symbolic emotional response to a much broader set of neighborhood characteristics than the crime-related news that is transmitted through these networks. During the 1970s, "law and order" (and, by extension, the fear of crime) became an effective catchword used by some politicians to exploit public fears concerning many social issues (Krisberg 1991:143). Richard Taub et al. (1981:104; see also 1984) have argued that residents are often willing to tolerate relatively high levels of crime as long as other aspects of community life are sufficiently gratifying. However, when the economic future of an area is uncertain (Taub et al. 1984) or when there are visible signs of incivilities and disorder in the neighborhood (Garafalo and Laub 1979; Skogan and Maxfield 1981; Lewis and Salem 1986), high reported levels of fear of crime may actually serve as an indicator of a more general concern that the area is out of control. Therefore, expressions of fear may represent in part general feelings of "urban unease" (Garafalo and Laub 1979) due to the perception of community dynamics that are publicly assumed to be related to crime.

Given the complex conceptual nature of fear, the development of valid and reliable indicators of the fear of crime presents some of the most difficult problems of measurement of all those discussed so far in this book. Unfortunatey, as we will see in the following section, many of the most widely used indicators of fear have led to a great deal of confusion and inconsistency within the literature.

Measuring the Fear of Crime

As noted in the preceding section, Fred DuBow and his colleagues (1979:2–6) have identified three conceptually distinct dimensions that have often been used interchangeably to study the fear of crime: values pertaining to the tolerance of crime, judgments concerning the risk of victimization, and emotional responses to this risk. In addition, some studies have focused on personal aspects of these dimensions, while others have framed their questions in terms of a more general referent.

Kenneth Ferraro and Randy LaGrange (1987) have utilized this typological framework to classify typical questions that have been used to measure the fear of crime (see Table 4–1). Of the examples included in this table, the personal judgment of safety indicator has had an important role in the development of the fear of crime literature. Not only is it included in the National Crime Survey, but the seminal work of Wesley Skogan and Michael Maxfield (1981), which will be discussed extensively later in this chapter, bases its analysis of fear on this measure. Although Skogan and Maxfield (1981:49) clearly consider fear to represent an emotional response to a stimulus, the structure of this survey item makes it impossible to separate that response from the perceived risk of victimization (Garafalo and Laub 1979). Since the perceived risk of crime is correlated only moderately with

Table 4–1
Commonly Used Measures of the Fear of Crime: Types of Perceptions

Values

General: Choose the single most serious domestic problem (from a list of ten) that you would like to see government do something about (Furstenberg 1971).

Personal: Are you personally concerned about becoming a victim of crime? (Jaehnig et al. 1981).

Judgments

General: Do you think that people in this neighborhood are sare inside their homes at night? (Clarke and Lewis 1982).

Personal: How safe do you feel or would you feel being out alone in your neighborhood at night? (Liska et al. 1982).

Emotional

General: I worry a great deal about the safety of my loved ones from crime and criminals (Lee 1982).

Personal: How afraid are you of becoming the victim of (sixteen different offenses) in your everyday life? (Warr and Stafford 1983).

Reprinted from "The Measurement of Fear of Crime" by Kenneth F. Ferraro and Randy La Grange in *Sociological Inquiry*, Vol. 57 (p. 78) 1987, by permission of the author and the University of Texas Press.

the emotional response to crime (see Ferraro and LaGrange 1987:79), the confounding of the two dimensions in a single indicator introduces a significant amount of measurement error into statistical models of fear.

In addition, Skogan and Maxfield argue (p. 50) that the phrase "being out alone in your neighborhood at night" orients the question to a fear of street crime that is likely to be committed by people from outside the household. However, as Garafalo observes, crime is not even mentioned in the question (much less particular forms of crime), leaving the issue implicit. Skogan notes in a later work (1990:76) that responses to this question may also reflect perceptions of neighborhood disorder that are not specifically criminal. Thus, in additoin to confounding perceived risk and emotional response, such safety-based questions confound crime and disorder, which are related but conceptually distinct.

Ferraro and LaGrange (1987:77) argue that survey questions that address the personal emotional response to crime through the notion of "being afraid" come much closer to providing valid and reliable indicators of fear. However, as noted in the preceding section, fear represents an emotional response to particular situational contingencies. Although studies that have examined the fearful reactions to such contingencies are relatively rare, there are some notable exceptions. Mark Warr (1990, for example, has examined the extent to which unfamiliar environments, darkness, and the presence of others provoke fears of criminal victimization. Adri van der Wurff and his colleagues (1989) also have investigated the roles of novelty and darkness, as well as the effects of contact with a potential aggressor, the presence or absence of a deviant behavior, and the existence of sexual connotations. Unfortunately, the neighborhood contexts of such situational contingencies have received very little attention.

A final problem with many measures of the fear of crime is that the nature of the potential victimization is often unspecified. For example, Ferraro and LaGrange (1987:77) note that while questions concerning "areas around here, i.e., within a mile of your home, where you would be afraid to walk alone at night" have emerged as standard indicators of the fear of crime, the nature of the perceived potential victimization that may lead to reported levels of fear is rarely made explicit. This is a critical consideration, for strong evidence has been presented by Warr and Stafford (1983) that the fear of crime is a multiplicative function of both the risk of victimization and the seriousness of the potential event. That is, their research clearly suggests that fear can only be understood fully within an offense-specific framework. For example, later work by Warr (1984) indicates that the relationship between perceived risk and fear is not consistent across sixteen different offenses.

The very general nature of the indicators typically used to measure the fear of crime raises a very important issue in the evaluation of this body of literature. On the one hand, perhaps the emotional reactions to all crime share enough common characteristics that the unique features of particular situations and types of potential victimizations can be safely ignored (see the discussion of J. Smith 1991). On the other hand, perhaps such "omnibus" approaches (Warr 1984) obscure the dynamics of this reaction more than they illuminate it.

To address this issue, we factor-analyzed the correlations among the levels of fear reported by young men and elderly women for the index offenses presented in Warr (1984:697; see Table 4–2). These two populations have generally been found to be the least and most fearful, respectively, in the United States. Despite their very different compositions, a single dimension was found to underlie the fear-related responses of both groups, thereby indicating that crime-specific perceptions tend to have a significant degree of shared variation. Therefore, while omnibus models of fear of crime cannot provide a full sense of the situational dynamics that may lead to such responses, they do appear to reflect a core emotional response that seems to be fairly generalizable across crimes. Nevertheless, due to the other measurement problems noted in this section, especially concerning the wording of the fear of crime survey items, the conclusions concerning the fear of crime that legitimately can be drawn from the research to be discussed in this chapter should be considered very carefully.

Table 4–2
Factor Structure of Reported Fear of Crime

	Young Men	Elderly Women
Robbery	0.796	0.910
Assault	0.698	0.904
Murder	0.733	0.921
Rape	—	0.927
Burlary while home	0.579	0.793
Auto Theft	0.467	0.278
Burglary while away	0.686	0.828

Based on data presented in War (1984). The rape item was not asked of male respondents.

Fear and the Threat of Crime: Neighborhoods and Indirect Victimization

It would seem to be logical to propose that those people with the highest risk of victimization or who have actually been victimized would exhibit significantly greater levels of fear. However, although some studies have confirmed the relationship between prior victimization and the fear of crime (see Skogan 1986), the magnitiude of this relationship is not as strong as one might suppose, even when adjustments are made for the degree to which people are exposed to risky situations (Stafford and Galle 1984). For example, while Skogan and Maxfield (1981:60–61) observe that robbery victims were more likely than nonvictims to report that they felt very unsafe in their neighborhoods, they also note that almost one-quarter of the nonvictims reported such emotions. In fact, the findings of some studies (such as Taylor et al. 1986; Greenberg 1986) suggest that the relationship between victimization and fear is essentially zero at the individual level.

Two general arguments have been proposed to account for this anomaly. The first suggests that certain groups perceive themselves as particularly vulnerable to crime in the sense that they are more open to attack, powerless to resist attack, and more exposed to traumatic physical and emotional consequences (Skogan and Maxfield 1981:69). As Mark Warr (1984:695; 1991) has argued, such considerations may lead to a "differential sensitivity to risk"; that is, some groups will exhibit higher levels of fear than others when in equally risky situations. Warr argues that this may be due to either differences in the perceived seriousness of particular offenses (since the seriousness of the activity affects the degree to which fear and perceived risk are related) or to the tendency of some groups to perceive certain clusters of illegal activities as being temporally coincident in particular criminal events. As Warr notes, "A high perceived probability of residential burglary may provoke intense fear among many women because assault, rape, and even homicide are viewed as likely contemporaneous offenses" (Warr 1984:695). In particular, some women perceive rape to be a logical precursor or outcome of a variety of offenses, such as burglary, robbery, or homicide (Warr 1985, 1991).

While the vulnerability hypothesis emphasizes particular individual-level characteristics that may heighten a person's sense of fear, the systemic structure of residential neighbohoods is the core element of a second explanation that has been offered to account for the paradoxical relationship between victimization and fear. Criminal events affect far more people than those directly experiencing the victimization. Rather, Taylor and Hale (1986:156) note that accounts of such experiences can spread throughout the local relational networks of a community, thereby creating a "shock wave" that spreads the impact of the victimization. Therefore, the degree to which one is embedded in local community networks affects the amount and nature of crime-related information to which one is exposed. Since people who listen to descriptions of crime may model the anxiety felt by those who actually have been victimized (Riggs and Kilpatrick 1990:131), exposure to such information may increase a person's fear of crime. That is, the fear of crime may reflect the vicarious or indirect effect of victimizations that have occurred to others.

One important source of information that directly may inflate the fear of crime is certainly the media, whose coverage of crime emphasizes violent activities (Skogan and Maxfield 1981:Chapter 8). Tom Tyler and Fay Lomax Cook (1984:693) have shown, however, that mass media presentations concerning crime and violence are generally unrelated to individual perceptions of the risk of victimization. Tyler (1984:34) offers two explanations for the lack of such a relationship. First, most citizens do not find such crime-related news particularly informative because the less spectacular crimes for which citizens are at the greatest risk (such as burglary and purse snatching) are significantly underreported relative to their rate of incidence. Second, news presentations concerning serious crime tend to concentrate on activities within high-crime areas, rather than providing a representative depiction of the distribution of crime within an entire city. As a result, Tyler and Cook (1984:694) argue that people develop their perceptions of the risk of victimization on the basis of their own experiences or, more important, those they learn about indirectly through their friends, co-workers, or neighbors. Thus, victimizations that occur outside one's extended network of relationships are unlikely to be given serious consideration when a person evaluates his or her risk of victimization and, by extension, should only be weakly related to the fear of crime.

Two basic processes are implicit to the indirect victimization model (Skogan and Maxfield 1981:147–152; see Figure 4–1). First, conversations concerning crime problems are most likely when residents perceive that

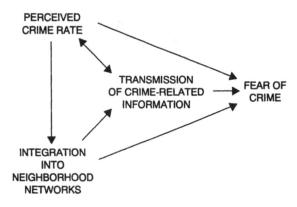

Figure 4–1. The Basic Indirect Victimization Model.

there are such problems in the neighborhood. Thus, the degree to which local networks transmit crime-related information is a function of the perceived frequency and seriousness of crime in the neighborhood. Second, the level of exposure to indirect sources of victimization is related to the degree to which residents are embedded in these networks.

It is worth highlighting an interesting implication of the indirect victimization model. We have argued throughout this book that the capacity for systemic control is greatest in those neighborhoods whose residents are highly integrated into relational networks. Yet, it is precisely these same networks that facilitate the transmission of crime-related information. Thus, as Skogan (1986:211) notes, those areas in which these stories can spread most widely are also those in which the levels of victimization tend to be fairly low (see also Furstenberg 1971; Skogan and Maxfield 1981).

The systemic implications of the indirect victimization model make it difficult to make direct predictions concerning the relationship between networks of relationships and the fear of crime. On the one hand, one might expect that those persons who are strongly integrated into their neighborhoods would be exposed to the greatest amounts of crime news and, as a result, would report the greatest fear of crime. On the other hand, neighborhoods with high levels of integration also have greater capacities for collective responses to crime, which may in turn reduce the fear of crime (Gates and Rohe 1987:426). Therefore, the role of the systemic structure in the indirect victimization model is not clear.

Unfortunately, the resolution of this issue is not straightforward, for very few studies have directly measured the central group processes of the indirect victimization model. For example, Taylor and Hale's (1986) test of the model includes no measures of the degree to which neighborhood residents have been exposed to reports of local crime through relational networks. Rather, they focus on the effect of whether or not the respondent had witnessed a street crime during the last six months. Since this variable is significantly related to the degree to which a person is embedded in local networks as well as the reported level of worry about street crime, Taylor and Hale (p. 173) argue that their findings suggest that "witnessed crime leads to crimes being shared with others . . . leading witnesses to inquire about local events from co-residents." However, their analysis does not provide a direct test of this proposition. Therefore, while their findings are congruent with the vicarious victimization thesis, they certainly are not conclusive.

A number of studies have focused on general perceptions of the neighborhood crime rates without restricting such perceptions to those activities which have been witnessed personally. Jeannette Covington and Ralph Taylor (1991), for example, find that respondents who have heard about the victimization of other neighborhood households have higher levels of fear. However, several studies have found that the relationship between indirect victimization and the fear of crime is not completely straightforward (see Lewis and Salem 1986:51). For example, Merry (1981:136) observes that the black residents of the housing project she studied learned the identities of both victims and offenders through their personal associations. However, the information networks of the Chinese residents transmitted news only about Chinese victims, thereby making it seem that they were the primary targets of a population with whom they had very little contact and familiarity. Thus, black residents integrated into informational networks had more of a sense of control than similarly integrated Chinese.

Merry concludes (p. 136) that the manner in which social networks "channel and block the flow of information about crime and criminals significantly influenc[es] the way each group perceived the dangers of the project." Therefore, the effect of integration on the fear of crime may be conditional on the nature of the information that is transmitted through personal networks. Such confounding influences may explain why Ronald Akers and his colleagues (1987) find only a very weak (although significant) relationship between personal fear of crime and the knowledge of neighborhood crime among elderly persons living in four communities.

Few studies have examined the full set of neighborhood and group dynamics involved in the transmission of information that are at the heart of the indirect victimization model. The most noteworthy exception is the work of Skogan and Maxfield (1981), which explicitly examines the degree to which the transmission of crime-related information is related to the fear of crime. While their work confirms the prediction of the model that persons who perceive greater problems of crime in their neighborhood are more likely to talk about this issue with others, perceived crime in itself does not guarantee that such discussions are disseminated throughout local networks.

Rather, those residents with strong residential and social ties to the community are significantly more likely to discuss these issues with other neighbors than those who are more weakly integrated into the area.

Skogan and Maxfield examine the effects of these patterns of communication through an analysis of the degree to which their respondents were aware of anyone in the neighborhood who had experienced a burglary, personal theft, assault by a stranger, or rape. Those respondents who talked to other neighbors about local crime problems were significantly more likely to know local victims. In turn, those who had such knowledge were significantly more likely to feel "very unsafe" in the community.

The findings of Skogan and Maxfield concerning the content of information transmitted through local networks have important implications for the evaluation of the body of research that has examined the indirect victimization hypothesis strictly on the basis of neighborhood integration. Albert Hunter and Terry Baumer (1982), for example, present evidence that people who are more likely to recognize strangers in their community and who feel like a part of the neighborhood are less likely to express a fear of crime.[4] Similarly, Covington and Taylor (1991) find that neighborhoods in which more respondents feel that neighbors would call the police if kids were spray-painting a building have significantly lower levels of fear. On the other hand, Timothy Hartnagel (1979) finds no relationship between neighborhood integration and the fear of crime[5] while studies based on the Greenberg Atlanta data discussed earlier in the book (such as Taylor and Hale 1986; Gates and Rohe 1987) report that highly integrated residents tend to exhibit higher levels of fear.

A key consideration in the resolution of the contradictory findings of these integration-based studies is the manner in which the variable has been mesaured. Skogan and Maxfield (as well as Hunter and Baumer 1982 and Lewis and Salem 1986) differentiate between social ties (the ease of identifying strangers and local juveniles in the neighborhood, and feelings of belonging) and residential ties (years in the neighborhood, home ownership, and residential expectations). While the Skogan and Maxfield research suggests that the relationship between integration per se and the fear of crime is primarily conditional, they note that the fear of crime has a weakly negative zero-order relationship with the level of social ties. The relationship is even more attenuated when residential ties are considered and is significantly confounded with age (Skogan and Maxfield 1981:116; see also Lewis and Salem 1986:54).

Unfortunately, the great variation in the measures of integration that have been used in the indirect victimization literature makes it very difficult to directly compare findings. The study of Hunter and Baumer utilizes measures of integration that are nearly identical to those used by Skogan and Maxfield and by Lewis and Salem. On the other hand, Taylor and Hale construct their measure of community integration on the basis of involvement in neighborhood activities, the likelihood of sharing information with neighbors, perceived similarity with neighbors, and the number of friends and relatives in the neighborhood. Likewise, the integration measure of Covington and Taylor reflects community responsiveness, those of Hartnagel reflect neighboring activities, and those of Gates and Rohe reflect perceived similarity and neighboring activities.

Such disparities in the operationalization of the key variables make it very difficult to reach definitive conclusions concerning the effects of indirect victimization. Given the fact that the fullest consideration of these dynamics (i.e., that of Skogan and Maxfield) provides support for the model, we feel that it is safe to assume that to some extent, the fear of crime is a reflection of the neighborhood context of indirect victimization. However, it is also apparent that these vicarious effects are not of such a magnitude that they can fully account for the paradoxes concerning the fear of crime that have been noted.

Fear as a Symbolic Response to Neighborhood Disorder

Although the indirect victimization perspective emphasizes the community context of the fear of crime, it still assumes that such fear is primarily an emotional response to the perceived likelihood of victimization, based on the combined experiences of oneself and others in the neighborhood. While we have presented some support for this model, it neglects the possibility that the fear of crime may represent in part a symbolic response to a wide range of neighborhood conditions that are not intrinsically crime-related. A second important approach has emphasized the tendency for residents to associate the likelihood of criminal victimization with certain community characteristics that are perceived to be related to a lack of local control in the area. This approach generally has been referred to as the social control or disorder model of fear (Lewis and Salem 1981, 1986; Greenberg et al. 1985; Skogan 1990; Taylor and Hale 1986).

The disorder model argues that fear is a response to the perception of residents that the area is becoming characterized by a growing number of signs of disorder and incivility (such as loitering groups of unsupervised teenagers, vandalism, graffiti, abandoned buildings, and public drug and alcohol use) that indicate that the social order of the neighborhood is eroding. Unlike the indirect victimization approach, which focuses on the community response to activities that are intrinsically threatening (such as learning of a rape or armed robbery in the neighborhood), the signs of disorder may not in themselves be especially frightening. However, they certainly symbolize such potential threats to many people (see Warr 1990:903).

The presence of graffiti in a neighborhood clearly illustrates the symbolic implications of disorder. There has been a great deal of debate in the national media concerning the relative status of graffiti as art or vandalism. Graffiti is simply the unauthorized use of public space as a medium for the presentation of messages and, in itself, the application of spray paint to the side of a building is a fairly innocuous, although sometimes irritating, behavior. However, graffiti may symbolize certain neighborhood behaviors that residents find very fearful, especially if the messages represent signs of gang activity or territoriality. These symbolic aspects were grapically illustrated recently in Norman, Oklahoma.[6] A resident of a neighborhood that was characterized by a growing concern over the potentially illegal behavior of its adolescents celebrated his birthday by inviting many of his friends to come to his new house and let their creative juices flow by using his backyard fence as an outlet for their artistic talents. Unfortunately, the principal of a local school drove by the yard the next day, noticed the messages and drawings that were displayed on the fence, and called a special meeting of the faculty to inform them that she had observed signs that gang activity was now firmly entrenched in the neighborhood.[7] To paraphrase Marshall McCluhan, the symbolic aspects of graffiti as a medium are much more important to many residents than the actual messages contained in those scrawlings.

The relationship between community dynamics and symbols of disorder were discussed extensively in Chapter 2, and there is no need to duplicate that presentation at this point. However, it may be worthwhile to emphasize the central arguments of this perspective. First, it is assumed that the presence of disorderly behaviors reflects the breakdown of accepted standards of public behavior (Lewis and Salem 1986:xiv). Second, it is assumed that disorder may be perceived by residents as leading directly to increases in crime (Wilson and Kelling 1982:31). Finally, the presence of disorder can lead to a breakdown of community cohesion as residents perceive that conditions in the neighborhood are getting out of control (Skogan 1990:47). The relationships underlying the disorder model are presented in Figure 4–2.

Recall that earlier in this chapter we noted that fear is an emotional response to a particular situational contingency. Therefore, even though signs of disorder may not in themselves represent the outcomes of illegal behavior (as in the case of groups of unsupervised youths), they can significantly increase the fear of crime in a neighborhood if residents perceived them as harbingers of impending danger (Stinchcombe et al. 1980:41). Lewis and Salem (1986:79) note that fear may particularly increase if residents no longer feel assured that:

1. Neighbors will adhere to a shared set of expectations about appropriate behavior

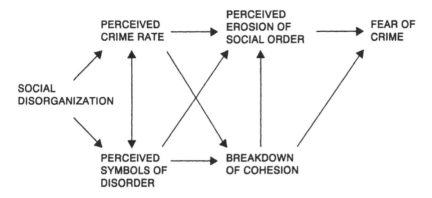

Figure 4–2. The Basic Disorder Model.

2. Private property will be kept up in accordance with commonly accepted standards
3. Public areas will be adequately maintained
4. Access to the community is regulated to control the influx of groups, businesses, and institutions that threaten the integrity of the area

Such community dynamics may significantly exacerbate existing feelings of personal vulnerability in three primary ways. First, the presence of loitering groups of youths is often associated with the verbal and sometimes physical harassment of people using local streets and facilities, especially women and the elderly (Skogan 1990:26–29). Second, symbols of disorder may represent novel social or physical aspects of a neighborhood whose meaning may not be clear to residents who have not experienced them in the past (Warr 1990). As we have argued, a familiarity with one's residential environment is a key element in the sense of control over potentially threatening situations. The emergence of symbols of disorder where none had been previously perceived therefore may undermine this sense of familiarity and lead to increased levels of fear. Skogan and Maxfield (1981:112–115), for example, note that perceptions of undesirable trends in neighborhood conditions are strongly related to the fear of crime.

Finally, and perhaps most important, the signs of actual impending criminal victimization are relatively sporadic and transitory during the course of a resident's neighborhood activities. However, as Arthur Stinchcombe and his colleagues note (1978), signs of community disorder are fairly enduring conditions that residents may be exposed to on an ongoing basis. Therefore, these symbolic cues may increase overall levels of anxiety by continually reminding residents of the possibility of future victimization.

As opposed to the findings that were presented for the indirect victimization perspective, the empirical literature provides strong evidence that the perception of high levels of disorder in a neighborhood is associated with a relatively high fear of crime (see, for example, Covington and Taylor 1991; Greenberg 1986; Lewis and Salem 1986; Skogan 1990; Taylor and Hale 1986). However, this generally strong conclusion must be tempered by three important considerations. First, the disorder model assumes that the levels of crime and disorder on the fear of crime may be highly confounded and very difficult to separate. While the correlation between fear and perceived disorder is 0.67 in Skogan's (1990:77) data, the correlations between perceived disorder and two measures of area crime conditions are 0.79 and 0.81; when the effects of the crime measures are taken into account, the relationship between disorder and fear disappears. However, Skogan argues that given the high interrelationship between these variables, a much larger sample would be necessary to fully untangle the association between disorder and fear. Such analyses confirm the independent effects of disorder (see Covington and Taylor 1991).

Second, Lewis and Salem (1986:100–101) present a strong argument that the relationship among crime, disorder, and the fear of crime depends on the perceived ability of local community organizations to influence municipal service bureaucracies to meet the needs of the area. For example, Richard Taub and his colleagues (1984:183) argue that corporations can be used to mobilize such bureaucracies, can create the "proper" normative climate for the promotion of a neighborhood, and can devote some of their resources to the community in such a way that the stability of the area can be increased. These efforts, in turn, may lead to a reduction in the fear of crime. Relatedly, DuBow and his associates (1979) note that collective responses to crime may actually reduce the fear of crime in an area regardless of any corresponding changes in the levels of crime or disorder. While we will address the issue of local community responses to crime in the final chapter of this book, it is worth noting at this point that such considerations have received very little attention in the evaluation of the association between disorder and the fear of crime although the perceived capacity of local governments to deal with local problems is intrinsic to at least one definition of the fear of crime (S. Smith 1989:198).

Fear as a Symbolic Response to Neighborhood Heterogeneity

One of the most visible and salient symbols of neighborhood status in the United States is the racial and ethnic composition of the area. Public opinion polls have generally documented an increasing trend in racial tolerance among American whites during the last few decades (see Schuman et al. 1985). In addition, Robert Lichter and his associates (1987) have discussed the significant changes that have characterized television portrayals of blacks (although notably *not* Hispanics) since the early 1950s. Nevertheless, a large concentration of blacks in a community is considered by many whites to be indicative of an undesirable neighborhood (Berry and Kasarda 1977:22), and crime control in heterogeneous neighborhoods often comes to be defined as "watching for people of particular races and aggressively monitoring the circumstances under which different races come into contact" (Skogan 1990:132). Therefore, it is not surprising that several studies (such as Lizotte and Bordua 1980; Liska et al. 1982; Moeller 1989) have observed that after controlling for the effects of crime rates and other relevant variables, the presence of racial minorities in a community is associated with relatively high reported levels of fear of crime among white residents. We will refer to this as the heterogeneity model (see Figure 4–3).

In part, these patterns of fear reflect the enduring public sterotype held by many whites that crime is primarily committed by members of minority groups (Swigert and Farrell 1976:2). This has especially important implications for the study of the fear of crime, for as Fishman et al. (1987) point out, sterotypes are created in part to predict how members of unfamiliar groups will behave. Thus the belief of many whites that minority groups have a higher propensity to crime would result in a perception that interaction with members of such groups is likely to lead to victimization.

Such beliefs are more prevalent than many whites would like to admit. For example, Reynolds Farley and his colleagues (1979) note that 59 percent of the white Detroit residents interviewed in their 1976 study felt that blacks were more prone to violence than whites. More recently, large proportions of the adult, white population of Oklahoma City have reported that on the average, blacks are more violent than whites (48.9 percent of the sample agreed with this statement), more involved with drugs than whites (60.8 percent), and more involved in criminal activity than whites (69.2 percent; Sonleitner et al. 1992). It might legitimately be argued that one cannot generalize the patterns of Detroit and Oklahoma City to the nation as a

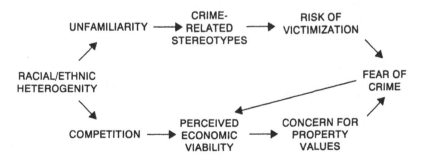

Figure 4–3. The Basic Heterogeneity Model.

whole. Yet 56 percent of the white respondents interviewed during the 1991 General Social Survey of the United States expressed the opinion that blacks are more violence prone than whites; 50 percent felt the same way about Latinos (Fulwood 1991).

While the bulk of empirical research has focused on the reactions of whites to the actual or potential presence of minority groups within their neighborhoods, there is good evidence that the fear of crime reported by members of minority groups is also a partial reflection of existing intergroup hostilities. For example, Merry (1981:126–143) reports that many of the Chinese residents of the housing project in which she conducted her study felt that blacks were predatory, violent, and singled Chinese out for victimization. In turn, many of the black residents described the Chinese in similar terms. Very similar group hostilities have been observed among the Mexican, Puerto Rican, Black, and Italian residents of the "Addams" area of Chicago (Suttles 1968) and among Mexican, Puerto Rican, Black, and Cuban residents of the Wicker Park Chicago neighborhood (Lewis and Salem 1986). More generally, the findings of Covington and Taylor (1991) indicate that the fear of crime is a declining function of racial and ethnic similarity of one's neighbors.

Owing to the growth in power of the New Right and religiously based social movements such as the Moral Majority during the Reagan era, social problems have come to be increasingly viewed as individual problems of weakness or immorality (Reinarman and Levine 1989:126–127). Thus, as Barry Krisberg (1991) has observed, crime has come to be identified with the activities of evil people. If members of competing racial and ethnic groups are perceived as being especially prone to crime and violence, it is a relatively small inferential step in today's social climate to conclude that members of such groups are inherently evil, thereby exacerbating the existing racial and ethnic schisms in our society. It is no accident that 48 percent of the white respondents interviewed in the Farley et al. (1979) study felt that blacks were less moral than whites. Thus, the actual or potential presence of rival groups within one's neighborhood not only may heighten one's sense of unfamiliarity with a formerly comfortable environment, but may also suggest to some residents that the area has been (or will be) invaded by "evil" populations.

Some sociologists, such as Craig Reinarman and Harry G. Levine (1989), argue that the images of violence that are often attributed to rival ethnic and racial groups are fostered by the manipulation of public perceptions of these groups. To the extent that this manipulation is deliberate, the heightened fears of crime that may result represent the outcome of an intentional attempt to represent these groups as threatening to a presumed social order. For example, despite the general positive trends in the portrayal of blacks on television noted by Lichter et al. (1987), some very popular shows have thrived on stereotypes of racial and ethnic minorities as criminal/

evil. Consider the case of "Miami Vice," one of the most successful prime-time police dramas of the mid-1980s. Ben Stein (1985) has noted that despite the fact that two of the three central characters were members of minority groups (black and Cuban), virtually *all* nonwhite characters outside of the station house were involved at least marginally in criminal activities. The portrayal of blacks on the show was especially troubling in that they did not apparently have the organizational skills necessary to be successful in large-scale crime (Stein 1985:42). What is more, during the first season the show often featured a jive-talking, cowardly black con man whose reactions to danger and facial expressions reminded some of the infamous Stepin Fetchit character of the 1930s.

At other times, the means by which the fear of particular racial and ethnic groups is transformed into the fear of crime are much subtler and most likely unintentional. For example, a recent series of stories in the local Norman, Oklahoma newspaper focused on a section of the city that was characterized by an apparently high level of criminal activity during a three-month period. In this series, a police officer working in Norman's Street Crimes Unit attributed the situation to an increase in gang-related activity in the area. Although he noted that the members of these gangs were Hispanic, white, and black, the entire series of news stories was dominated by ongoing references to the Bloods and Crips, infamous (and increasingly apocryphal) black gangs based in Los Angeles. Not only is "Crips" graffiti prominently displayed in a front-page photograph, but the only anecdote in the series that contains any kind of ethnic identification pertains to an incident that occurred between a white woman and four black men.

Thus, despite the comment of the officer concerning the heterogenous racial composition of the suspected offenders, some Norman residents have come to identify the problem with the low-income black residents of the area. The degree to which this seemingly innocent series of articles heightened racial hostilities in the area can be illustrated dramatically by an incident that occurred shortly after the stories appeared in the paper. A local reporter was called by an acquaintance who lived in the area, who stated that she had just seen a black man get out of a car that had parked down the block from her home. She wondered if she should call the police department to report such "suspicious activity."

Criminologists must share part of the blame for the widespread public association of crime with minority status. Thanks to numerous public statements and press releases that have been issued, the public is well aware that many studies have found a strong positive correlation between the minority composition of a neighborhood and the crime rate. However, throughout the preceding chapters of this book we have noted many studies that have found that the relationship between racial composition and crime disappears when the effects of other variables (such as the dominant family structure in the neighborhood) are controlled. Such findings are rarely emphasized by criminologists in high-visibility public settings and, therefore, are not part of the general public consciousness concerning crime.

The evidence that has been presented suggests that incidents of victimization are often reinterpreted and redefined within the context of existing intergroup hostilities (Merry 1981:143). As such, the fear of crime is partly a specific realization of a more general fear of unfamiliar racial/ethnic groups that one may find threatening because of their "strangeness." This approach to the fear of crime may seem to be primarily a criminological version of the contact hypothesis, which assumes that stereotypes lose their relevance when social interaction increases among once-hostile groups. However, in addition to hostilities that may arise through unfamiliarity, conflict may also arise among groups competing for the same set of scarce resources; such competition was apparent among the groups studied by Merry despite the fact that all were residents of a public housing project (see Chapter 5). In fact, the traditional contact model argues that interaction will only lead to a decrease in prejudicial stereotypes when those processes occur among people with equal status in cooperative circumstances. Thus, while familiarity with the residents of one's neighborhood is certainly a key dimension underlying the fear of crime, contact alone cannot lead to a decline in crime-related prejudice if the involved groups perceive that they are competing for some scarce resource.

An interesting competition-based alternative to the contact perspective has been offered by Richard Taub et al. (1984). While they acknowledge (p. 181) that racial and ethnic stereotypes may serve as symbols of crime and community deterioration, they argue that the primary symbolic effect of racial composition pertains to the perceived economic viability of a residential area. In part, their argument concerning the fear of crime is grounded in an indirect victimization model, for they assume (p. 168) that it is primarily a function of personal or contextual (i.e., vicarious) victimization. However the resulting level of fear may or may not be considered an acceptable cost of living in a particular community, depending on the perceived dangers of living in alternative environments and the perceived values of rewards and amenities the person receives on the basis of his residential location. As the disorder model would suggest, the greater the level of neighborhood deterioration and crime, the greater the dissatisfaction with the neighborhood.

On the basis of this model, Taub et al. present findings that suggest that those people most dissatisfied with an area have the greatest fear of racial change (p. 177). However, this is not necessarily due to a perceived inherent criminality on the part of particular racial and ethnic groups. Rather, increasing crime, deterioration, and dissatisfaction may lead to an exodus of white residents, an increase in the demand for local housing by blacks, and a decrease in the competitiveness of the housing market (p. 179). Thus, the Taub et al. model argues that the relationship between fear of crime and hostility toward blacks among whites primarily reflects perceived potential economic losses in this market.

The primary implication of the Taub et al. research is that in areas with relatively high rates of crime, deterioration, and dissatisfaction, the fear of crime may lead to a fear of racial change that will be associated with further economic decline. Since this may lead to residential relocation, these fears may therefore lead to a further erosion of community stability and increased neighborhood deterioration. That is, as opposed to the models that have been discussed to this point, the fear of crime is not only an outcome of community dynamics but is also a determinant of those processes. We turn our attention to this issue in the final section of this chapter.

The Fear of Crime as a Source of Neighborhood Change

In Chapter 2 we discussed the degree to which criminal behavior can significantly shape the systemic character of urban communities. Similar observations have often been made concerning the fear of crime. While high levels of anxiety about crime may lead individuals to avoid certain locations in the neighborhood or to take protective measures, they may also lead to formal and informal collective action directed against the source of that fear (Lavrakas and Herz 1982; Krohn and Kennedy 1985; Gates and Rohe 1987). Harvey Krohn and Leslie Kennedy (1985:698) have identified three basic types of such action: *coproduction*, which represents interactions between individuals or groups and public agencies with the goal of augmenting or contributing to the provision of urban services pertaining to local levels of crime: *ancillary production*, which represents individual contributions to the delivery of public services oriented to crime prevention; and *parallel production*, which represents individual activities that do not involve a public body. Given the arguments that we have made concerning neighborhood participation and systemic control, such activities might be assumed to lead to increasing levels of community integration. That is, the fear of crime would lead to an increase of local solidarity.[8]

On the other hand, three counterarguments can be made to this position. First, there is good evidence that the fear of crime in itself may not be a primary motivating factor that leads to involvement in local activities. Paul Lavrakas and Elicia Herz (1982), for example, find that the fear of crime does not significantly differentiate between residents who do and do not participate in neighborhood crime prevention programs. Rather, such participation usually was an outgrowth of a person's involvement in other broadly based community groups that were not oriented specifically toward crime prevention per se. Similarly, Timothy Hartnagel (1979) concludes that there is no significant association between the perception of changes in local levels of crime and neighborhood cohesion.

Another possibility is that high levels of fear of crime actually erode people's willingness to take positive action against crime, either individually or collectively (Skogan 1990:67). Finally, Hartnagel (1979) and Skogan (1990) both suggest that the effect may be curvilinear. That is, while an organized response to crime is not likely in neighborhoods in which the fear of crime is very low, it is also not likely in the most fearful areas in which "demoralization and distrust prevail" (Skogan 1990:69). However, moderate levels of fear may lead to such organization and a resultant increase in integration.

Unfortunately, it is impossible to evaluate the relative accuracy of such hypotheses on the basis of the available research. The simultaneous estimation of the reciprocal relationship between fear and community organization necessitates the use of fairly complex nonrecursive statistical models. In fact, just as we noted in Chapter 2 in the case of crime, if fear and organization are reciprocally related, then models that calculate only the effects of one variable on another (i.e., either the degree to which organization affects fear or the degree to which fear affects organization) have produced biased estimates of those effects.

There have been some efforts to estimate nonrecursive models of the fear of crime. D. Garth Taylor et al. (1986), for example, examine a system of equations in which fear and defensive actions (such as walking in groups, the avoidance of public transportation, and the installation of home security devices) are assumed to have simultaneous effects on one another (see also Liska et al. 1988). Likewise, Allen Liska and Barbara Warner (1991) have examined the relationship between city crime rates and changes in personal activities. However, both these studies focus on individual reactions to fear, and the research of Liska and Warner is framed at the city, rather than local community, level. We are unaware of any research that has used such an approach to examine the relationship between the fear of crime and various indicators of the systemic organizational structure of residential neighborhoods.

Conclusions

For over twenty years, criminologists have grappled with the sometimes paradoxical and puzzling patterns of crime-related fear that characterize residents of the United States. To some degree, these efforts have been hampered by inconsistent definitions and measures of the fear of crime, which sometimes has resulted in apparently contradictory findings. Nevertheless, it has become clear that while fear represents a response to the perceived likelihood of victimization to some degree, it also has clearly symbolic dimensions that reflect more general perceptions of residential neighborhoods. Although a great deal of valuable work has been conducted, we are not yet close to a complete understanding of how this complex emotional response is related to local community dynamics. As was the case in each of the previous chapters, the difficulty of collecting reliable and valid data pertaining to the central hypothesized dynamics is one of the most important shortcomings of this body of research. Nevertheless, work in this area has provided a rich set of insights into the neighborhood and crime relationship that are highly complementary to those discussed in the preceding chapters.

5

The Neighborhood Context of
Gang Behavior

> I very much fear that the gang has been the theoretician's Rorschach
> in criminology—one can easily find what he [sic] seeks
> —(Klein 1971:viii).

Malcolm Klein's observation concerning the role of street gang research in the development of criminological theory is hard to dispute, for many of the current etiological theories of crime have deep roots in this tradition. For example, the development of control and social disorganization theory was strongly influenced by the gang work of Frederic Thrasher (1927), and Clifford Shaw, Henry McKay, and associates (1929, 1942); the strain and cultural conflict theories by the research of Albert Cohen (1955), Herbert Block and Arthur Neiderhoffer (1958), Richard Cloward and Lloyd Ohlin (1960), and by certain arguments of Shaw and McKay (1942); economic conflict theory by William F. Whyte (1981) and William Chambliss (1973); and learning theories by the general body of literature on group dynamics and crime. Thus, many criminologists appear to have agreed at least implicitly with Klein's earlier (1969:64) statement that gang behavior is "certainly the most interesting form of delinquency."

Despite the important theoretical impact that gang research has had on criminology, widespread interest in the dynamics of such behavior has not been a consistent feature of the discipline. In the introductory chapter to his influential *Street Gangs and Street Workers*, Klein (1971:1, 4) states simply that "I've had it with gangs." Klein immediately noted that this sentiment did not mean that there was nothing left to be learned from gang research. Rather, it reflected his perception of diminishing personal rewards from such work, since "once you've seen 300 gang meetings, you've seen them all" (p. 4). Such sentiments certainly are understandable given the extensive field research that Klein had conducted with gangs in Los Angeles. However, it appears that his statements were more than a little prophetic, for after a period of prominence that peaked during the 1960s, gang research seemed to fall dramatically on the list of criminological priorities.

Now, in the closing years of the twentieth century, gangs again have emerged as a widely discussed social concern, not only among academics and criminal justice personnel but among the general public. Many schools have forbidden students to wear certain items of clothing that are perceived by administrators as connoting gang membership. This concern is not restricted to the largest metropolitan areas of the country. We recently attended a school board meeting in a small, rural community in central Oklahoma during which a revision of the dress code was discussed extensively. While the principal of the local junior high school acknowledged that the area did not yet have a "gang problem," she argued that "authorities" had informed her that if the proper steps were not taken (one of which was the imposition of a strict dress code), the area would be overrun by gang activity within five years.

It might be assumed that this renewed interest is a response to qualitative changes that have occurred in the nature of gang activity during the last decade. For example, Irving Spergel's (1990) superb review of the gang literature suggests that contemporary gangs may be more violent (especially with the increased availability of semi- and fully automatic assault weapons), more involved in the use and distribution of drugs, older, and more prevalent in smaller communities than was the case in the past. Joan Moore (1988:4) warns against broad generalizations, for the nature of gang activity may vary significantly within a single city. Yet even if such changes have occurred, they are not sufficient to account for the revitalization of gang research. For example, although many people were under the impression that gangs no longer represented a major problem in urban areas during the 1970s, Walter Miller (1975) provided evidence that gang activity was more violent than it had been in the past. Therefore, the renewed interest in gang crime is a reflection of a broader set of dynamics than simply the character of contemporary activity.

As we observed earlier in this book, criminology has been characterized by ongoing paradigm shifts concerning the role of group dynamics in the etiology of crime. Hedy Bookin and Ruth Horowitz (1983) have argued that the apparent "end of the youth gang" primarily was a function of the ideological dominance of individual-level theories of crime and delinquency that were especially influential during the 1970s. As a result, very little attention was paid to the group dynamics of crime that had been emphasized in gang research. However, the pendulum of attention has again shifted and gangs have regained a place of prominence among academic topics. Perhaps one of the most interesting indicators of this revival is that Malcolm Klein is again one of the most prolific contributors to the current body of literature.

An enormous number of topics could be addressed in a treatise on gang behavior. Given the purposes of this book, we have restricted our attention to the neighborhood context of such behavior.[1] However, perhaps more so than in the preceding chapters, it is especially important to clarify our conceptualization of the gang and gang crime, for definitional difficulties have led to a great deal of confusion, debate, and consternation among those who have tried to develop widespread generalizations concerning the current status of these phenomena.

Preliminary Concerns

Defining Gangs and Gang Behavior

Peggy Sanday (1990) has provided a detailed description of the dynamics that led to a gang rape alleged to have been committed by members of a fairly well-organized, cohesive group of older adolescents in Philadelphia. Prior to this particular incident, the XYZs (a fictitious name) had already developed a widespread reputation in the neighborhood for problematic behavior. Women commonly reported that they had been verbally harassed by members of the gang who hung around drinking beer on benches along the primary street in the area. Since these benches were situated in front of their clubhouse, the group made it clear that this was their "turf" to do with as they pleased.

Although all the members of this gang were enrolled in school, the group allocated some degree of special status to those who performed poorly. One of the judges involved in the rape case noted that a statement that the XYZs had offered into evidence was "ungrammatical . . . replete with misspellings . . . garbled and incomprehensible" (Forer 1990:xvii). As Sanday has reported, new members of the community were commonly warned about the group, and women were urged to consider the potential dangers of attending the parties that were regularly thrown by the gang.

To many, this short description has all the hallmarks of classic, popular descriptions of a gang, that is, a group of inner-city adolescents, a concern with turf, harassment of local residents, an organizational structure, some degree of solidarity, and mutual participation in serious forms of illegal behavior. Sanday notes (p. 71) that during her two interviews with one of the people implicated in the gang rape, "[H]is dislike for what I was doing and his sense of superiority to people like me were expressed throughout. . . ." In general, we would guess that most readers would not consider this to be a group with which they would like to interact on a regular basis. However, we have left one very important piece of information out of our short summary of Sanday's study: these were all members of a prominent fraternity at a prestigious, upper-middle-class university; the neighborhood in question was a campus community in Philadelphia.

Perhaps some readers think that this is an inappropriate example of gang activity, for public images of such behavior usually do not include the activities of fairly affluent fraternity members at highly respected colleges.

Yet consider the influential definition of a gang provided by Klein (1971:13): any identifiable group of youngsters who (a) are generally perceived as a distinct aggregation by others in their neighborhood, (b) recognize themselves as a denotable group (almost invariably with a group name), and (c) have been involved in a sufficient number of delinquent incidents to call forth a consistent negative response from neighborhood residents and/or law enforcement agencies. As anyone familiar with campus life is aware, all fraternities easily qualify under the first two conditions; each has a unique name, and highly visible, relatively arcane symbols (i.e., Greek letters) are used to signify membership in such groups. The third condition is the one that would disqualify many (and perhaps most) fraternities. Yet Sanday's ethnographic material clearly shows that the "XYZ" fraternity had a "dangerous" reputation on campus, and we would be surprised to find many college campuses without at least one such house. Nevertheless, despite the fact that the XYZs clearly qualified as a gang under Klein's definition and the fact that one of the judges described the similarities of this case to those involving more traditional gang members as "striking" (Forer 1990:xvi), there is no indication that Philadelphia's long-established Gang Crimes Unit had any involvement in the case.

Thus, we come to the heart of the problem: exactly how are we to define a gang? Without a generally accepted definition of the concept, it is impossible to make any kind of informed judgment concerning the nature and extent of gang behavior, much less changes that have occured over time. Some criminologists would certainly include the XYZ case in the computation of rates of gang crimes[2]; others would object strongly to such a classification.

Likewise, what are we to make of the fact that many small, stable, rural communities have recently claimed to be the site of gang behavior? For example, a Knight-Ridder newspaper item (Wallace 1991) describes the case of Frederick, Oklahoma (population 5,200), where the local police chief believes that violent, drug-dealing gangs are staking out territory in the community. The primary basis for his conclusion is the existence of some "Bloods" graffiti in the area, a number of auto thefts, cases of shoplifting and intimidation, reports of drug dealing, and warning notes (this time from the Crips) that have been left on cars.

It is clear that one of the gravest mistakes that a community can make is to deny the existence of a gang-related problem until a series of serious incidents force the issue. Columbus, Ohio, for example, denied that it had any type of problem until members of the mayor's family were brutally attacked by people claiming gang affiliation (Huff 1989:530–531). Therefore, the situation in Frederick might be seen as an outcome of the expansion of gang-controlled drug markets that had been widely discussed (U.S. General Accounting Office 1989). However, while the concern of the Frederick police chief certainly is understandable, a healthy degree of skepticism is warranted concerning the large number of communities who suddenly have discovered a "gang problem" in their midst. Grasmick was recently told by a high school teacher from Oklahoma City (which does have a documented gang problem), that once particular symbols (such as certain forms of dress or graffiti) became associated with gang membership in his school, they quickly became adopted by many nongang adolescents as a sign of personal rebellion.[3] Therefore, the incidents that were reported in Frederick (including the graffiti) may not be gang-related in any respect other than they represent the efforts of local youths to adopt symbols that are guaranteed to elicit a horrified reaction from the adults in their community.

There are other dynamics that also must be considered when evaluating the extent to which an area is characterized by gang activity. Many concerned communities have invited law enforcement personnel to speak to local leaders about whether they have a gang problem and, if so, what they should do about it. One of the central themes that usually emerges is that without proper action on the part of the community, it is likely to be overrun with gang-related problems in a relatively short period of time. Such messages are quickly picked up by the local media and spread to the general public. Hagedorn (1988:30), for example, reports that the elites and the media of Milwaukee adopted an image of gang behavior that was promoted by the Chicago Gang Crimes Unit and reinforced by "scary slide shows of murders and a display of gang weapons that would make the U.S. Army run for cover." During these presentations, Milwaukee was warned that if the city failed to act "in a hard line manner," Milwaukee's gangs would be like those in Chicago within five years or less. We find this passage to be especially interesting in that one member of Oklahoma City's Gang Crimes Unit has worked especially hard to promote the image of impending gang danger across the state. Recall that the principal of the junior high that was quoted in the beginning of this chapter reported that she was told that if her community took no action, gang problems would be rampant within five years.[4]

Finally, it must be noted that since access to some federally funded law enforcement programs is more likely if a gang problem can be demonstrated in a community, some agencies may have vested interests in the "discovery" of gang activity. It is impossible to determine the extent to which the apparent diffusion of gang behavior reflects such economic considerations. However, such dynamics have been suggested as an explanation of why police estimates of the number of gangs in Phoenix increased from 5 or 6 to over 100 in a very short period of time (Zatz 1987).

It certainly has not been our intention in this section to downplay the seriousness of some gang activities or to imply that most communities are exaggerating the problems they face. Rather, we have attempted to emphasize that without a precise and parsimonious understanding of what constitutes a gang and gang behavior, it is often difficult to separate fact from mythology. Unfortunately, several factors make it very difficult to arrive at such an understanding.

Crime and Delinquency as Group Phenomena

There are very few issues concerning which criminologists usually feel confident enough to make strong declarative statements. However, the group nature of delinquency is certainly one of those issues. One of the most influential findings of the Shaw and McKay research was that almost 90 percent of the delinquent events reflected in the juvenile court records of Cook County involved two or more participants (Shaw et al. 1929:7–8). The group orientation was strongly reinforced several years later when Edwin Sutherland (1934) began to develop his influential theory of differential association, which emphasized the small-group dynamics associated with the learning of delinquent and criminal behavior. More recent work has noted some important offense-specific differences in the rates of group offending. In addition, a large proportion of offenders do not engage in illegal behavior strictly in group situations (see the review of Reiss 1988). Nevertheless, the presumed group nature of illegal behavior is a generally uncontested part of criminological lore.

There have been several important criticisms of the group hypothesis. The differential association perspective suggests that the most important sources of information concerning the techniques, motivations, and justifications for illegal behavior are intimate personal groups (see Sutherland's propositions 3 and 4). Given the apparent group nature of crime and delinquency, the intimate nature of these groups might suggest that offenses occur primarily within aggregations with temporal histories, fairly developed sets of relationships among the members, and relatively high levels of cohesiveness and solidarity. However, Klein (1969) has argued that the existence of two or more offenders in a single incident does not in itself guarantee that the event represents the outcome of such group dynamics. He criticizes in particular the influence that the Shaw and McKay findings have had on the discipline, for they were based on official records in which it is impossible to determine the actual group dynamics that may have been involved.

Klein illustrates this problem with several hypothetical examples, including one in which a relatively large number of strangers are attending a party and they happen to purchase marijuana from one of the other attendees. If the police happen to bust the party and make multiple arrests for possession, the arrest reports would most likely note that several people were involved in the incident. However, they would not constitute a group in any sociological sense of the word. Rather, these people were simply "contiguous individuals" (Klein 1969:67) who were engaged in the same behavior in the same location. Because of such conceptual ambiguities, some researchers now utilize alternative phrases (such as "co-offending"; Reiss 1988) to refer to events in which more than one person was involved but in which the existence of group dynamics is not clear.[5]

Klein certainly is not arguing that group dynamics are unrelated to criminal and delinquency behavior. Rather, he is emphasizing the need to recognize the basic distinction between the sociological notions of aggregate and group processes. In that respect, some unknown percentage of illegal behavior may be more validly viewed as a form of collective behavior in which an aggregate of relative strangers respond to a particular stimulus; this aggregate may have a very limited prior history and may disband after that particular response. There is a large body of literature that indicates that persons are more likely to engage in illegal behavior if their closest friends are involved in such behavior (see Elliott et al. 1985). However, there also is evidence that many delinquent behaviors occur in the company of individuals to whom a person has relatively weak associational bonds. Martin Gold (1970:83–94) has likened this situation to a "pickup game" of basketball in which the roster of players depends on who happens to be on the playground at the same time. That is, those present may define an opportunity as suitable for basketball (delinquency), and once the game is concluded, many of them go their separate ways. While certain interesting dynamics are involved in the definition of the situation, they have a relatively short-term relevance to the participants. These are not typically the kinds of processes that sociologists attribute to groups. As a result, some criminologists have raised important questions concerning the extent to which group solidarity is reflected in the illegal behavior of co-offenders (see Morash 1983).

Much of the confusion that has arisen in the gang literature, as well as in the public's perception of gang behavior, is due to the often interchangeable use of the words "group" and "gang" (see the criticism of Klein and Maxson 1989). For example, while Walter Miller (1980) delineates twenty different types of "law-violating youth groups," he only considers three of these to represent gangs (see Table 5–1). If the general pattern of relationships among co-offenders is much more fluid than is usually assumed, perhaps the primary distinction between group and gang crime and delinquency pertains to the internal dynamics of the aggregate that may result in a criminal event. For example, Bernard Cohen (1969:66) considers delinquent groups to represent relatively small cliques that coalesce sporadically without apparent reason and spontaneously violate the law. Cohen considers such groups to be ephemeral, with no elaborate organizational structure, name, or sense of turf. Gangs, on the other hand, are highly developed aggregates with relatively large memberships. As opposed to delinquent groups, gangs have elaborate organizations, names, senses of corporate identity, and identifications with particular territories. A similar typology

Table 5–1
Types and Subtypes of Law-Violating Youth Groups

1	Turf gangs
2	Regularly associating disruptive local groups/crowds
3	Solidary disruptive local cliques
4	Casual disruptive local cliques
5	Gain-oriented gangs/extended networks
6	Looting groups/crowds
7	Established gain-oriented cliques/limited networks
7.1	Burglary rings
7.2	Robbery bands
7.3	Larceny cliques and networks
7.4	Extortion cliques
7.5	Drug-dealing cliques and networks
7.6	Fraudulent gain cliques
8	Casual gain-oriented cliques
9	Fighting gangs
10	Assaultive cliques and crowds
10.1	Assaultive affiliation cliques
10.2	Assaultive public-gathering crowds
11	Recurrently active assaultive cliques
12	Casual assaultive cliques

Reprinted from Miller (1980) by permission of the editors.

has been developed by Irving Spergel (1984). The viability of such distinctions will be examined in the next section.

Defining Gang Delinquency

John Hagedorn (1988) has identified two primary ways in which the gang has been defined within the criminological literature. The first, and oldest, approach has emphasized the processes that give rise to such groups. Albert Cohen (1955), for example, defines gangs in terms of collective reactions to problems of social status, while Richard Cloward and Lloyd Ohlin (1960) focus on the interaction between legitimate and illegitimate opportunity structures. However, we feel that the most important processual definition for understanding the relationship between neighborhood dynamics and gang behavior is that of Frederic Thrasher (1927), who defines a gang as "an interstitial group originally formed spontaneously and then integrated through conflict. . . . The result of this collective behavior is the development of tradition, unreflective internal structure, esprit de corps, solidarity, morale, group awareness, and attachment to a local territory" (p.46).

Several aspects of Thrasher's definition are worth noting. First, "interstitial" has a dual connotation. Thrasher uses it in one sense to represent the period of life when one is neither a child nor an adult; gangs therefore are a reflection of the period of adjustment between childhood and maturity (p. 32). For this reason, Thrasher argues that such groups are relatively short-lived and that adult gangs or members are relatively rare.[6] Yet this does not mean that gangs are characterized by age homogeneity. Rather, as older members age out of the group, younger members join, leading to a set of loosely connected, age-based cliques within the gang.

Thrasher also used the term "interstitial" to refer to neighborhoods located between Chicago's central business district and "the better residential areas" (p. 6). Since these were areas characterized by neighborhood deterioration and residential turnover (p. 46), Thrasher's model is clearly a variation of the social disorganization approach that we have discussed extensively. The systemic implications of his approach are clearly evident in his discussion (p. 33) of the failure of "directing and controlling customs and institutions to function efficiently in the boy's experience." The spirit of this aspect of Thrasher's processual definition is evident in Spergel's (1984:201) more recent definition of integrated gangs as a reflection of the inability of primary and secondary community institutions to provide mechanisms of opportunity or control.

Second, Thrasher's emphasis on "spontaneous formation" reflects his argument that all childhood play groups represent potential forms of gangs (see pp. 23–26). Since such groups usually arise on the basis of interaction and familiarity, they tend to form around particular residential locations in a neighborhood where youths are likely to come into contact with one another. Thus, street corner groups represent the basic building block upon which Thrasher develops his thesis. The key determinant of the transition into a gang is contact with other groups (either other play groups or adults) who express disapproval or opposition to the playgroup. For example, Hagedorn (1988:57–60) observes that fierce rivalries developed among the many breakdancing groups that arose during the early 1980s in Milwaukee; gangs sometimes emerged as a result of the fights that often broke out after competitions. Such conflict can produce an awareness of the distinction between "us" and "them" and the development of a sense of solidarity among group members. The existence of a street corner group therefore can serve as a source of protection from other groups in the neighborhood (see Spergel 1984: 202).

Finally, note that delinquent or criminal activities are not mentioned in Thrasher's definition. While he certainly recognized that such activities may be facilitated by gang membership, he emphasized the variability that existed in the 1,313 groups that he identified as gangs: some are good, some are bad. Thrasher's approach emphasizes the social dynamics that may lead to cohesion among a play group and the resulting development of a gang. The relationship of gangs to delinquency is therefore a key analytical issue.

Although such process-based definitions of the gang continue to appear in the literature, Hagedorn (1988:57) notes that most current research is no longer characterized by a focus on how gangs arise within particular community contexts and how they function within those social environments. Rather, he argues that the fundamental question has become "why gang members are delinquent." The definition of Klein (1971) presented earlier in this chapter, with its criterion that the number of delinquencies committed by the group has called forth some type of negative response, represents a commonly used example of such an approach. The implications of this shift in focus are much more important than they may first appear, for illegal behavior is considered to be a definitional aspect of gang activity, whereas for Thrasher it was an empirical question.

Even more so than was the case with processual definitions, there is an enormous variety of delinquency-based definitions which have become the basis for different policies, laws, and strategies. One of the most interesting attempts to produce a definition with a broad consensual base is that of Walter Miller (1975, 1980), who asked a national sample of youth service agency staff members to respond to the questions: "What is your conception of a gang? Exactly how would you define it?" His final definition is based on the responses of 309 respondents representing 121 youth serving agencies in 26 areas of the country (Miller 1980:120), including police officers, prosecutors, defenders, educators, city council members, state legislators, ex-prisoners, and past and present members of gangs and groups (1980:117).

Of the 1,400 definitional characteristics that were provided by his sample, Miller reports that there were six items with which at least 85 percent of the respondents agreed (1980:121): a youth gang is a self-formed association of peers, bound together by mutual interests, with identifiable leadership, well-developed lines of authority, and other organizational features, who act in concert to achieve a specific purpose or purposes, which generally include the conduct of illegal activity and control over a particular territory, facility, or type of enterprise.

Such delinquency-based definitions have been criticized for several reasons. Klein and Maxson (1989:205) call Miller's approach "discouraging" and argue that to define a concept on the basis of the results of a "vote" does not make it inherently more definitive or valid than other approaches. Yet their criticisms are not aimed solely at Miller, for they note that the definitional task is "difficult and arbitrary" and an "inherently unsatisfying task." The continued existence of a great variety of delinquency-based definitions (see Spergel 1990) suggests that consensus does not exist for any particular conceptualization (although the definition provided in Klein 1971 has been particularly influential). Ruth Horowitz (1990:43) notes that the variation in locally used definitions may be useful for understanding how the relationships among criminal justice personnel, the community, the gang, and the individual gang member are defined. Nevertheless, the lack of a standard, nationwide definition of a gang makes estimates that have been made concerning the number of youth gangs in the United States or comparisons that have been made over time or between communities relatively meaningless (Spergel 1990:180).

Some contemporary researchers have expressed a more general discomfort with all definitions that assume generalizable groups structures and processes or that equate crime with gang behavior (see Hagedorn 1988; Fagan 1989:643). Merry Morash (1983:310) argues that these approaches developed due to a growing reliance on definitions used by law enforcement and social work personnel. Since many of these agencies classify groups as gangs if violent or criminal activity is a major activity, gangs are by definition heavily involved in illegal behavior and Thrasher's question concerning the relationship between gang membership and delinquency becomes tautological (see Short 1990:160).

To illustrate the implications of such definitional assumptions, Morash created a scale of "gang likeness" based on an adaptation of Miller's definition.[7] While her analysis presents evidence that the gang-likeness variable has a significant effect on delinquency, more general peer group processes, such as the delinquent behavior of one's friends, are of much greater importance. Overall, she concludes (p. 325) that membership in a stereotyped gang is not a sufficient condition to stimulate serious delinquency. This seems to provide an important contradiction to the finding of many studies that gang members are involved in significantly higher levels of crime and drug use (see Fagan 1989). However, Klein and Maxson (1989:204) take issue with Morash's findings, noting that adolescent church or school groups could have qualified as gangs using her criteria.

Other characteristics of gangs that might be the subject of empirical investigation are also embedded into definitions such as that developed by Miller. For example, while some of the informants in Hagedorn's (1988) study reported that their gangs had fairly specialized and formalized ranks, others insisted that the structure was very informal; a few even stated that their gangs had no recognized leader (p. 92). Likewise, Joan Moore (1978:44) reports that the historical circumstances that set the context for the development of each of the age-based cliques (*klikas*) in Los Angeles Chicano gangs has resulted in significant differences among groupings in the same gang, each of which may have its own organizational structure (see also Keiser 1969:15).

We find the arguments of Hagedorn and Morash very persuasive, for those characteristics that are assumed by researchers such as Miller to be defining features of gangs actually exhibit a great deal of variation among groups who have been identified as gangs. Rather than taking these characteristics for granted, it would seem to be much more theoretically fruitful to examine the processes that give rise to such group variation. Perhaps one of the reasons why the Klein (1971) definition has been extremely popular is that the three criteria are extremely flexible and are relevant to a wide range of gang types.

Nevertheless, we are uncomfortable with the delinquent behavior criterion, for it makes a possible outcome of gang activity one of the defining characteristics. Klein and Maxson (1989:204) defend their position by noting that "to think of modern street gangs independent of their criminal involvement is to ignore the very factor that makes them qualitatively different from other groups of young people."[8] Despite our own misgivings concerning the presumed equivalence of gang activity and crime, there is no question that the major criterion used by many audiences in the definition of gang is the group's participation in illegal behavior (Spergel 1990:179).

The Collection of Data for Gang Research

Throughout this book we have continually stressed the major difficulties that exist in collecting the data necessary for testing various hypotheses concerning neighborhoods and crime. This problem is especially acute when conducting research on street gangs.

Easily the longest tradition of gang research is based on some variant of ethnographic fieldwork with gang members (or the combination of such research with supplementary forms of data collection). The work of Thrasher (1927) is exemplary in this respect. Although we know very little concerning how he actually collected his data (see Short 1963:xviii), it is clear that it represented primarily a combination of personal observation and documents that were supplemented by court records and census materials. Over the course of his seven-year study, he amassed enough material to identify 1,313 Chicago gangs.

While Bookin and Horowitz (1983) noted that fieldwork techniques had a declining popularity in sociology and predicted that they would rarely be used in future research, Horowitz (1990:37) recently has retracted that statement, for they certainly represent one of the major forms of data collection used in the study of gangs.[9] Unfortunately, ethnographers no longer have the resources at their disposal to conduct such a "census" of gangs as that of Thrasher. Therefore, the modern emphasis has been on the depth of data, rather than the breadth. Generally this is not considered a problem in fieldwork, for such research is much more concerned with the identification and analysis of process and meaning than with the ability to generalize findings to some larger population. Nevertheless, it must be emphasized that the representativeness of the gangs that have been described is not clear. Many times the gangs have been chosen on the basis of their notoriety within a community (see Keiser 1969; Muehlbauer and Dodder 1983), because of chance circumstances that bring a gang to the attention of a researcher, such as the prior participation of gang members in social service projects (Short and Strodtbeck 1965; Klein 1971; Hagedorn 1988; Harris 1988), or because of their location in particular communities upon which researchers have elected to focus (Klein 1971; Moore 1978; Horowitz 1983; Campbell 1984; Sullivan 1989; Jankowski 1991).

There are special difficulties in conducting fieldwork with gangs that do not arise in many other fields of inquiry. First, and most obviously, while most researchers are highly educated, middle-class persons, many gang members are not. It takes a skilled ethnographer to overcome the initial hostility that is often inherent to interactions with gang members (see the descriptions provided by Horowitz 1983:Chapter 1; Moore 1978:Appendix A; Hagedorn 1988:32–33). In addition to this inherent suspicion, many researchers have noted that gang members are notoriously unreliable as informants (Spergel 1990:175); Klein (1971:18) feels that "the only thing worse than the young reporter's description of a gang incident is his [sic] acceptance of the gang participant's statement about it." This problem was forcibly driven home to the first author of this book during a conversation with a friend who formerly had been a central member of one of Chicago's most notorious fighting gangs. He described with great pleasure how during times of boredom, members of his group would have an informal competition to see who could convincingly tell the most outrageous story to a social worker who had been assigned to work with the group. Therefore, the collection of valid data through fieldwork with gangs is only possible after an extended period of contact during which trust is established.

There is also a more subtle problem in the reliability of data drawn from fieldwork with gangs. An important concern in all ethnographic studies is the degree to which the presence of the researcher has a significant effect on the nature of the dynamics that are observed. For example, some of the most important studies of gang dynamics (such as that of Short and Strodtbeck 1965) have relied to a significant degree on the observations of "detached workers," that is, social service personnel who have been assigned to work with gangs in their natural settings. As Klein (1971:151) notes, those procedures that are often used to maximize contact with gangs (such as group counseling sessions or attendance at club meetings) may in fact increase group cohesiveness, which may lead to an increase in gang delinquency. In addition, the assignment of a group worker may increase the local reputation of a gang, which in turn may attract new members. Thus, the presence of a fieldworker can result in a set of group dynamics and activities that would not have occurred otherwise.

Despite these problems, ethnographic work has provided some of the most important insights that criminologists have about gangs, and much of the richest data has been obtained under situations that may have seemed doomed to failure (see Horowitz 1983). However, as we have noted, there are problems in the generalizability of such data. A second approach to gang research has attempted to overcome this limitation by incorporating surveys into the study design. While this often has been done in conjunction with ongoing fieldwork (such as Short and Strodtbeck 1965; Joe and Robinson 1980), this is not necessarily the case (Giordano 1978; Bowker et al. 1980; Morash 1983; Fagan 1989; Curry and Spergel 1991).

Many of the same problems concerning trust and hostility that characterize fieldwork studies also are present in survey-based study designs. However, two other issues make gang survey research especially problematic. The first is the sampling frame itself, that is, the population of gang members from which the respondents should be selected. Obviously, there is no "official" listing of all gang members in an area, but even if one existed, the ongoing flux in gang membership would make a list obsolete almost immediately (Short and Strodtbeck 1965:10). The police in many communities have compiled lists of suspected gang members, but these tend to be very inaccurate. Klein (1971:19) tells the story of how he examined the files kept by the police concerning the members of a particular gang. Whereas he had the names of over 100 members, the police had less than 20 and much of their information concerning addresses and offense histories was extremely dated. One solution is to administer surveys to those people who have been identified as gang members through fieldwork (see Short and Strodtbeck). Another is to interview people known to be gang members, ask them for the names of other people who should be interviewed, and continue to build the sample of respondents through such a "snowball" approach (see Fagan 1989).

While such techniques can potentially collect a great deal of useful information concerning the characteristics and behavior of gang members, it is often desirable to compare the distributions of these variables to those found among youths not involved in gang activity.[10] However, the selection of an appropriate comparison group is very difficult. The sample survey data examined in the 1989 paper of Jeffrey Fagan (1989), for example, included only the responses of gang members, and he was only able to draw comparisons with nongang youths by comparing his findings with other published research. While Short and Strodtbeck (1965) did include nongang youths in their sample, all these respondents were affiliated in some manner with youth-serving agencies (p. 5). Therefore, the degree to which these youths are representative of nongang youths in general is not clear.

Several attempts have been made to identify gang membership and make the relevant comparisons on the basis of more broadly administered surveys (Morash 1983; Rand 1987; Spergel and Curry 1988; Curry and Spergel 1991). The validity of the information that has been collected on the basis of such study designs depends on two crucial considerations. First, how likely is it that youths involved in gangs will be represented in the sample?

Some researchers have tried to maximize this possibility by drawing all or part of their sample from those youths residing in correctional facilities (Bowker et al. 1980; Morash 1983). Such approaches would tend to over-represent those youths with extensive or especially violent offense histories. Other sampling designs are likely to underrepresent active gang members. Spergel and Curry (1988) and Curry and Spergel (1991), for example, sur-veyed all male students in the sixth through eighth grades at four schools in Chicago. Likewise, Fagan (1990) supplemented the gang data noted ear-lier with information collected from a sample of high school students re-siding in the same three neighborhoods and a snowball sample of dropouts. However, school-based samples are especially prone to errors in studies of delinquency since the most active delinquents may be those youths who are most likely to be truant during the time of administration. In general, it is extremely difficult to draw a representative sample of gang members.

The second consideration reflects the identification of respondents as gang members. Some surveys have simply asked the respondents if they belong to a gang (Rand 1987; Johnstone 1981). While John Johnstone presents some evidence (p. 362) to suggest that the adolescents in his sample interpreted the term "gang" consistently, he does note problems with such an assumption. Other researchers assume the existence of a continuum along with a youth group is more or less like a particular operational definition of a gang (Morash 1983; Spergel and Curry 1988; Curry and Spergel 1991). We have already noted Klein and Maxson's (1989) criticism of the scale developed by Morash for its apparent inability to differentiate among dra-matically different types of youth groups. The Spergel and Curry scale is a much narrower approach to the measurement of gangs and includes such items as the flashing of gang signs, the wearing of colors, and attacking (or being attacked) in a gang-related incident. One of their most important findings is that the indicators of gang involvement scaled differently for Hispanics and African-Americans, which highlights our argument that the search for a broadly relevant uniform definition of gangs may be relatively fruitless. In addition, contrary to the findings of Morash, they present evi-dence of a strong relationship between gang involvement and serious delin-quency.

Overall, the use of surveys is no guarantee that the results of a study are any more reliable than those produced through more traditional field-work approaches. Rather, results are especially sensitive to the nature of the sampling design, the selection and wording of the indicators of gang membership, and the relevance of those indicators to the populations under consideration.

The final technique that has been used to collect data on gangs is based on information that has been collected by law enforcement agencies. While sometimes this information is used to supplement that derived through fieldwork or survey designs, much of the current knowledge concerning gangs is the result of studies that have been based primarily on such data (Miller 1975; Spergel 1984, 1986; Curry and Spergel 1988; Klein and Maxson 1989; Maxson et al. 1985). Bernard Cohen (1969) has argued that the Philadelphia Gang Crimes Unit uses sociologically sophisticated definitions of gang and nongang activities in its classification of criminal events. However, the official classification of an offender as a gang member generally is not systematic and may not be based on reliable criteria (Klein and Maxson 1989:206).

In addition to the problem of identifying gang membership based on the information included in official records, there is an equally difficult problem in the classification of illegal events as gang-related. For example, suppose a member of a gang is arrested for the armed robbery of a convenience store. On the basis of the description of the event provided in the arrest report, it may be impossible to determine whether it was committed due to gang membership. Unfortunately, there are no national standards for the identification of a crime as gang-related. For example, the Los Angeles Police and Sheriff's Departments designate a homicide as gang-related "if either the assailant or the victim is a gang member or, failing clear identification, elements of the event, such as motive, garb, characteristic gang behavior, or attribution by witnesses, indicate the likelihood of gang involvement" (Klein and Maxson 1989:206–207). However, the Chicago Police Department uses a much more restrictive definition that is based on the evidence of "gang function or motivation" (Curry and Spergel 1988:384). Maxson and Klein (1990) note that a reclassification of the Los Angeles data on the basis of the Chicago criterion leads to a significant reduction in the estimated rate of gang homicide and question whether the massive efforts of gang control and suppression that have characterized Los Angeles would have developed if this alternative definition had been used to gauge the extent of the problem.

The existence of such definitional inconsistencies makes it very difficult to make any kind of reliable comparisons between jurisdictions concerning the level of gang activity. However, definitions may also be characterized by inconsistencies even within the same jurisdiction. For example, Curry and Spergel (1988:385) note that prior to 1986, arson, theft, burglary, and vice offenses (including those that were drug-related) were not included in the gang crime reporting system. Such changes make it nearly impossible to examine the trends in many forms of gang behavior in Chicago over time, including the changing nature of drug use and distribution that has received so much attention in other parts of the country.

Overall, the inherent limitations of the dominant forms of data collection on gangs are very serious. Therefore, in many respects, we simply cannot be as confident of our knowledge concerning gangs as we are in other areas of criminology. However, despite these problems of measurement, certain patterns have emerged in a sufficient number of studies and locations to provide at least a minimal degree of confidence in those empirical regularities. This is especially the case in gang research that has emphasized the neighborhood dynamics related to such behavior.

The Effect of Neighborhood Dynamics on Gang Behavior

Criminologists have identified two particular types of neighborhoods in which the development of gangs is most likely. It is ofen assumed that the most prevalent form of gang activity arises in unstable, institutionally weak lower-class neighborhoods. However, gang activity also has been commonly noted in relatively stable lower-class areas that are fairly well organized (Spergel 1984:203).

Three classes of theories have developed to account for these phenomena. While a systemic approach grounded in the theory of social disorganization has been the traditional explanation for the existence of gangs in the unstable areas, the presence of gangs in the second type of neighborhood usually has been accounted for either in terms of subcultures that support such behavior or, more recently, in terms of economic marginalization. The existence of such disparate approaches presents a very difficult logical problem for the development of a systemic theory of neighborhood crime, for the systemic model assumes that a general consensus exists concerning the negative evaluation of certain types of crime, whereas subcultural approaches assume that some neighborhoods may encourage such activity. Likewise, while economic factors have a primarily indirect effect on crime in a systemic model, they are at the heart of the marginalization thesis.

While it is possible that different explanations are required for different neighborhoods, it would be much more parsimonious if a single model was able to account for these situations. In this section we will examine the three frameworks, evaluate their relative validity in terms of recent empirical evidence, and examine the ability of a systemic approach to account for such findings.

The Traditional Systemic Social Disorganization Approach

Although Thrasher did not develop a systematic series of hypotheses on the basis of his data (see the criticism of Short 1963:xxi–xxii), all the elements of a systemic social disorganization model of neighborhood crime are reflected in his treatise. First, as noted earlier, he observed that delinquent gangs are most likely to arise in relatively poor, unstable neighborhoods. Such characteristics decrease the ability of local customs and institutions to control the leisure time activities of boys, which increases the likelihood of gang conflict (Thrasher 1927:33). While he considers the family to be the most important agency of control, he also discusses the inability of other institutions to restrain such behavior. Thus, all three forms of systemic control that were discussed in Chapter 1 are key elements of Thrasher's model (see pp.33, 65): the private (the family), the parochial (local churches and schools), and the public (corruption and indifference in local politics and exclusion from high-paying occupations).

Several large-scale studies have provided at least partial support for the social disorganization gang approach. Desmond Cartwright and Kenneth Howard (1966) examined the neighborhood characteristics of the sixteen primary gangs who were studied in the Short and Strodtbeck (1965) research. Based on the observation that these communities are "oversupplied with young children of both sexes and undersupplied with mature adults" (p. 342), they argue that the controls that adults can impose on youth in such areas are relatively weak. However, while they did find that gangs were more likely in low-income areas, they did not find a relationship between residential mobility and the distribution of such groups, which is an important contradiction of Thrasher's model. To some extent, this may represent the fact that gangs are sometimes found in stable, low-income neighborhoods, an issue that we will address in the upcoming sections of this chapter. However, it is important to emphasize that there is a serious measurement problem in the study since the sixteen gangs examined by Short and Strodtbeck represented neither an exhaustive nor a representative sample of those in Chicago at that time (see Short and Strodtbeck 1965:Chapter 1). Therefore, some unknown proportion of the neighborhoods that are considered to be "nongang" by Cartwright and Howard are actually characterized by the presence of such groups.

More recently, Irving Spergel (1984:201–202; see also 1986) has argued that the social disorganization model may provide the best account for the development of violent gangs. However, he differentiates between two different sets of community dynamics (or, in Spergel's terms, "routes") that could give rise to such behavior. The integrated route leads to the development of violent gangs in areas newly settled by lower-class populations. While in the past these groups tended to be Irish, Italian, Jewish, Polish, German, or African-American, the contemporary immigrants are primarily Hispanic or Asian. While families in these areas may be intact, "secondary institutions . . . are weakly identified or structurally connected with the interests and needs of the population" (p. 202). Within this context, street corner groups provide a form of mutual protection.

Spergel's integrated route clearly involves the kind of systemic processes that we have discussed. However, gangs may also arise through a segmented route in relatively stable, well-organized areas with somewhat effective local institutions. Nevertheless, these communities are very poor, economically dependent, have high rates of family instability, and have been abandoned by white populations. Gangs provide a number of illegitimate opportunities in such areas and have "more of an economic-gain character" (p. 203). Spergel illustrates the complexity of these two sets of dynamics in an analysis of gang homicides in Chicago between 1978 and 1985.[11] He finds, for example, that with one exception, extremely poor, black neighborhoods with very high delinquency rates tend not to have especially high rates of gang homicides. That is, as the dynamics of the segmented route suggest, violent gang behavior is rare in such neighborhoods.

In a later paper, David Curry and Spergel (1988) utilize multivariate techniques to examine more fully the empirical relationship between community dynamics and violent gang behavior. Two of their many findings are especially important. First, after controlling for the effects of poverty, the percentage of Hispanics in a neighborhood is significantly related to the rate of gang homicide. Second, poverty is related to gang homicide in primarily black and white neighborhoods, but not in Hispanic communities. To some extent, these findings support Spergel's distinction between integrated and segmented routes to gang violence. Given the history of settlement in Chicago, Curry and Spergel (p. 387) consider the concentration of Hispanics in a neighborhood to be a "simple and gross" indicator of social disorganization (unfortunately, they have no direct measure of neighborhood stability or institutional effectiveness). To the degree that this is a valid indicator of disorganization, the high levels of gang homicide in Hispanic communities regardless of the level of poverty in those areas supports the systemic model. At the same time, the relationship between poverty and homicide in black and white neighborhoods suggests that chronic economic deprivation is also associated with gang violence.

The existence of gangs in stable neighborhoods (which has been noted by many researchers; see Whyte 1981) may appear to be an important contradiction of the systemic model and, in fact, has been a serious stumbling block for traditional social disorganization approaches. Part of the difficulty of the systemic model in this respect can be traced to the ongoing influence of early models of human ecology (such as Park and Burgess 1924; Burgess 1925), which assumed that immigrant groups eventually would be assimilated into the occupational structure of urban areas. Since residential upgrading would accompany economic mobility, these groups would relocate into progressively better neighborhoods with greater levels of systemic control. Mobility was assumed to lead to a great degree of contact between groups, which in turn would result in a decreasing relevance of racial or ethnic membership (Park and Burgess 1924:Chapter XI).

As a result, it was assumed that the racial and ethnic character of gangs would change over time. For example, because of their status as recent immigrant groups, the major street gangs that have been identified in the early nineteenth century in the United States tended to be primarily Irish (Haskins 1974:22). Even early Chinese immigrant communities, which have sometimes had a public image of being relatively gang free, gave rise to tongs, which often were involved in illegal activities (Haskins 1974:58). As the older immigrant groups became assimilated, they were replaced in their previous neighborhoods by newer arrivals, who then dominated neighbhood gang activity.

This idealized version of invasion and succession was modified by three important features of urban life. First, as we noted in Chapter 1, some racial and ethnic groups were formally or informally prevented from engaging in the residential mobility that was assumed to accompany economic assimilation. As a result, some neighborhoods have mixtures of socioeconomic groups. For example, Brigitte Erbe (1975) has presented evidence that in Chicago during 1970, black professionals and managers lived in neighborhoods with occupational compositions comparable to those areas where unskilled white workers lived. Relatedly, because of their marginal economic position, certain extremely poor groups have very few prospects for residential mobility and may be "abandoned" in particular neighborhoods for several generations. Finally, certain ethnic groups have been characterized by a reluctance to move from the "old neighborhood" even when it was economically feasible to do so. The discussion of Horowitz (1983:56) suggests that this may be due to a reliance on the expanded family for emotional and social support; she notes that even when people move away from the neighborhood, intensive relationships are often maintained.

These urban dynamics necessitate several modifications of Thrasher's argument. Most notably, his thesis suggests that the protective functions of street corner groups would decline in importance in neighborhoods with extensive formal and informal ties among the residents. Given the presumed residential mobility of racial and ethnic groups, gangs should be found primarily among the most recent groups of immigrants to an urban area. To some extent this is true. An increasing amount of attention has been paid to the apparent rise of Asian youth gangs (see Joe and Robinson 1980; Spergel 1990:212–213, 216; Chin 1990; Vigil and Yun 1990), and some parts of the United States are characterized by ongoing Hispanic in-migration, which has been associated with the development of gangs. However, Hispanic gangs have been a feature of the American Southwest for over fifty years, dating at least to the zoot suit riots during World War II (see Moore 1978). Likewise, many black gangs have exended histories, and Spergel (1990:213) notes the continued existence of gangs among second- and third-generation Italians, Irish, Polish, and middle Europeans. Therefore, gangs are not a transitory part of the urban experience of many racial and ethnic groups.

It is worth noting that the common public image is that gangs are a predominantly Hispanic and African-American phenomenon, especially since these groups currently constitute the largest numbers of youths arrested for gang offenses (Spergel 1990:213). To some extent, the apparent concentration of gang crime in these groups is to be expected given their respective histories of urban seclusion and economic marginality. However, although the data are now fairly dated, it is important to note the research of Leonard Savitz and his associates (1980), which indicates that 12 percent of black youths and 14 percent of white youths living in Philadelphia during the mid-1970s claimed gang affiliation. Similar distributional findings also have been reported by Mark Testa (1988; in Spergel 1990). Thus, blacks and Hispanics have no special predisposition to gang membership. Rather, they simply are overrepresented in those areas most likely to lead to gang activity.

Unfortunately, many media presentations give the impression either that gang activity is fairly rare in other groups or that the nature of such activity is relatively minor.[12] Such impressions are false on both counts. Seven of the thirty-seven gangs studied by Martin Sanchez Jankowski between 1978 and 1989 had a membership that was strictly white, and even in the cities in which gang behavior is predominantly black/Hispanic (such as Milwaukee; see Hagedorn 1988), a number of white gangs exist. In addition, white gangs account for their own share of seriously violent behavior, as shown in a study of a white gang located in an affluent Chicago suburb: two of the members were sentenced to six years in prison for bombing the cars of a police chief and a vice-principal (Muehlbauer and Dodder 1983).

Contemporary urban dynamics necessitate several important modifications of traditional systemic analyses of neighborhoods and gangs. First, the nature of neighborhood territoriality has changed significantly. If entire groups are geographically mobile (as assumed by Park and Burgess), then the ethnic-based gang associations might be expected to dissipate as members become dispersed throughout the city. However, when residential areas retain their racial/ethnic identity over an extended period of time, those who migrate away may retain an affinity for the "old neighborhood." While some former members join gangs that exist in their new areas, others may retain their affiliation and possibly attempt to recruit new members (see Moore et al. 1983). Thus, while the residential neighborhood may serve as the original basis for the emergence of the gang, the mobility of gang members may expand the geographic range of the group.

The prison experiences of some gang members is a second way in which a group's territorial influence may expand beyond the boundaries of the residential neighborhood, for broadly based networks and alliances often develop during incarceration (Moore 1978). At the same time, neighborhood allegiances are often heightened during these periods of separation (Moore 1978:106–109). In at least one documented case involving the Vice Lords of Chicago, the correctional experience itself led to the formation of the gang in the neighborhood after the release of several members from the Illinois State Training School for Boys (Keiser 1969; Dawley 1992).

The apparent increase in the involvement of gangs in the distribution and use of drugs is a third factor that has been suggested as an explanation for the expansion of gang influence beyond neighborhood boundaries since the relatively quick profits that can be made has encouraged some gangs to expand their markets outside of their traditional areas (see Klein and Maxson 1985; Fagan 1990). For example, some law enforcement agencies have estimated that the Bloods and the Crips, two gangs based in Los Angeles, control approximately 30 percent of the national market in crack cocaine (U.S. General Accounting Office 1989). We have used the word "apparent" because while the traditional gang literature tended to downplay the role of drugs in urban gangs, Moore (1978) notes that the use of drugs such as barbiturates could be commonly found in Los Angeles barrios as early as 1940. Again, owing to inconsistencies and changes over time in the definition of gang activity, it is impossible to confidently document trends in the distribution of drugs by gangs. For example, recall that a drug-related gang incident was definitionally impossible in the statistics collected by the Chicago Police Department until 1986.

However, presumably because of an increasing involvement in the distribution of drugs, some observers have noted a tendency for gangs to expand their sphere of influence not only into new neighborhoods, but also into new cities. Sometimes this is simply due to the residential movement of gang members and not a planned, organized expansion of activities (Huff 1989). At other times there appear to be efforts to recruit within nearby cities. However, even when gangs affiliate and identify with larger "supergangs" in other cities, most of these linkages are fairly weak and members emphasize their relative independence from these large groups. Overall, Hagedorn (1988:77) concludes that despite the apparent influence of gangs from Chicago, most Milwaukee gangs originally developed in the traditional manner described by Thrasher. This does not imply that large-city networks of gangs do not exist. Rather this represents another gang-related issue in which it is very difficult to separate fact from mythology.

Finally, Hagedorn (1988:Chapter 6) has developed the provocative argument that the desegregation of public schools has contributed to the decline of gangs as a neighborhood-based phenomenon. While over 80 percent of the black gang leaders indicated that most gangs originally had a neighborhood basis, members were bussed to public schools all over the city. In addition to increasing the alienation of these youths from local institutions such as neighborhood schools, gang members were thrown into educational environments that contained gang members from many different parts of the city. As a result, they were forced to recruit members from these schools "not based on neighborhood but on the need for protection" (p. 136). The implications of such programs on the neighborhood basis of gangs can be clearly observed in Milwaukee since its desegregation plan involved mandatory busing for only blacks. While Hagedorn (p. 137) notes that less than a third of black gang members felt that the residential area of a recruit was an important consideration, the neighborhood was an important consideration for Hispanic and white gangs.

In summary, although the traditional systemic social disorganization approach continues to be a viable explanation of the relationship between neighborhoods and gangs, it is incomplete in two respects. First, such models traditionally have had trouble accounting for the existence of gang activity in relatively stable, low-income neighborhoods. Second, certain urban dynamics that could not be foreseen by Thrasher have broadened the sphere of influence of some gangs beyond the boundaries of the residential neighborhood.

Gangs as the Carriers of Criminal and Delinquent Subcultures

The inability of the traditional systemic models of Thrasher and Shaw and McKay to account for the existence of stable neighborhoods with extensive histories of gang behavior led to the development of a very different orientation to the neighborhood and crime. While the literature contains several important variations of this perspective, all these theories hypothesize that such behavior reflects the existence of community subcultures that support gang activity.

Like the theory of social disorganization, the subcultural approach has a long criminological history, as illustrated in a passage first published in 1929: "Resistance on the part of the community to delinquent and criminal behavior is low and such behavior is tolerated and may even become accepted and approved. . . . Delinquent and criminal patterns arise and are transmitted socially just as any other cultural and social pattern is transmitted." The authors of this incipient statement concerning cultural transmission are none other than Clifford Shaw and his associates (1929:205–206), a fact that has caused a great deal of consternation among social disorganization theorists (see especially Kornhauser 1978). The subcultural elements receive even greater emphasis in their later work, in which it is stated that "within the same community, theft may be defined as right and proper in some groups and immoral, improper, and undesirable in others" (Shaw and McKay 1969:171). The primary role of the gang within such neighborhoods is the transmission of these values to its members and the fostering of behaviors that reflect these values.

Each of the three statements of the subcultural perspective that had especially important effects on the direction of gang research during the 1960s and early 1970s assumed that lower-class communities were characterized by a distinctive set of norms and values that fostered such behavior.[13] Walter Miller (1958) provided perhaps the purest cultural perspective in his discussion of "focal concerns," that is, "areas or issues which command widespread and persistent attention and a high degree of emotional involvement" (p. 7). Lower-class communities are assumed to have long-standing cultural traditions in which trouble, toughness, smartness, excitement, fate, and autonomy are important themes; in addition, lower-class boys are concerned with issues of belonging and status. Delinquent behavior, from this perspective, represents an attempt to achieve those standards of conduct.

The gang plays a central role in the daily life of boys who are raised within such communities. Miller argues that since male adults are absent from most lower-class households, or at best participate sporadically in household affairs, most lower-class families are directed by women. Therefore, he argues that belonging to a gang in which all the members are male provides "the first real opportunity to learn essential aspects of the male role in the context of peers facing similar problems of sex-role identification."

A very different subcultural approach to gangs is found in the work of Albert Cohen (1955) and Richard Cloward and Lloyd Ohlin (1960). Whereas Miller considers the subcultural features of lower-class communities to be an explicit set of norms and values with an ongoing integrity, Cohen and Cloward and Ohlin argue that these features arise in response to a rejection of particular features of the dominant middle-class culture. Since this rejection is assumed to be due to frustrations that arise when limitations in the opportunities provided to members of the lower class make it difficult to attain the dominant goals of the middle class, these two theories combine cultural and structural dynamics, making them "mixed" models of gang behavior (see Kornhauser 1978).

Cohen considers the dominant goal of all individuals to be that of status, which is generally defined as a positive recognition of an individual's personal and social attributes (p. 27). In the American society, such attributions are made on the basis of the extent to which a person strives, "by dint of rational, ascetic, self-discipline and independent activity, to achieve in wordly affairs. . . . the presumption [is] that 'success' is itself a sign of the exercise of these moral qualities." Cohen argues that the primary societal context for the acquisition of such positive recognition among adolescents is the school.

This presents certain difficulties for working-class youths.[14] Schools are staffed by middle-class individuals who are expected to foster middle-class personalities by rewarding middle-class expectations (pp. 113–114). However, although "most working class Americans are under the spell of this particular set of norms" (p. 87), they tend to be perceived as failures by their teachers owing to "their relative lack of training in order and discipline, their lack of interest in intellectual achievement, and their lack of reinforcement by the home in conformity to the requirements of the school" (p.115). Given the devaluation of the working-class status in a middle-class institution, this can lead to the development of a deep sense of shame.

The basic process that leads to the formation of gangs in this context is a change of reference on the part of the adolescent from an inherently frustrating pursuit of middle-class goals to the pursuit of goals that appear to be attainable. For this "reaction formation" to be an adequate solution, working-class adolescents must convince themselves that the original goals are not worth pursuing (pp. 53–54). Since most working-class youths were originally socialized into middle-class norms, this reaction must be quite pronounced, leading to an intense hostility to the norms of middle-class society (p. 133). As a result, Cohen argues that the delinquent behavior of working-class boys represents a reversal of middle-class norms: it is non-utilitarian, malicious, negativistic, and is characterized by an emphasis on short-run hedonism and group autonomy (pp. 25, 30).

Gangs arise through the interaction of a number of adolescents who face similar problems of status attainment and who evolve similar modes of adaptation (p. 59). This interaction leads to the development of "group standards" that reflect this frame of reference and the emergence of a sub-culture (p.65). As long as the gang continues to serve the status needs of the age groups that succeed the original founders of the group, it will continue to exist, thereby becoming a subcultural delinquent tradition within the local community that passes on knowledge, beliefs, values, codes, tastes, and prejudices to new members (pp. 12–13).[15]

The relationship between frustrated goals and the formation of gangs is also the central theme in the work of Cloward and Ohlin. However, their work focuses on a much more limited source of status: the achievement of economic success. In one respect, the framework of Cloward and Ohlin was greatly influenced by the classic work of Robert Merton (1938), which assumed that innovative (i.e., deviant or criminal) techniques of attaining economic success are likely to be developed when the conventional insti-tutional means (such as educational or occupational opportunities) are un-available. If a person blames the lack of success on individual shortcomings, such innovations will most likely occur as solitary activities. However, if the sources of frustration are perceived as the result of social injustice, the first acts of deviance reflect tentative steps toward the development of norms that will put the person at odds with carriers of the dominant conventional culture. Cloward and Ohlin argue that at this stage the deviant needs "all of the encouragement and reassurance he can muster" to defend his position and finds this by searching for others who are in a similar position (p. 126). Thus, the neighborhood gang is born.

This focus on the societal distribution of legitimate opportunities for economic success is a relatively straightforward, although important, ap-plication of Merton's theory of deviance. The primary contribution of Clo-ward and Ohlin to the analysis of neighborhoods and gangs is their extension of that theory to include a consideration of the illegitimate opportunities that are available to a youth in a local community. The nature of these opportunities depends on two general features of the neighborhood. The first is the relationship that exists between "immature and sophisticated offenders" in an area (p. 153). Cloward and Ohlin suggest that the transmission of criminal traditions and the recruitment of young adolescents into illegal activities is more likely in areas where different age groups are integrated through stable and intimate associations. In addition, one must also consider the degree to which those people engaged in ongoing patterns of illegal behavior are integrated with conventional persons in a neighborhood to form a "single, stable structure which organizes and patterns the life of the community" (p. 156).

Cloward and Ohlin draw heavily from the work of Solomon Kobrin (1951) in their discussion of the implications of such integration on the development of gang behavior. Kobrin (p. 657) notes the existence of two polar types of communities. In some areas, there is a systematic and organized integration of illegal and conventional life-styles. For example the leaders of illegal enterprises may participate in such local institutions as churches, social groups, and political organizations. Thus, conventional and criminal value systems have a "reciprocal" (p. 658) relationship in which they reinforce one another; in such areas, delinquency arises within at least a "partial framework of social controls" and there are "effective limits of permissible activity" (p. 658). At the other extreme are those communities in which the adult criminal activity is unsystematic and disorganized and the conventional and criminal value systems are in opposition to one another. Therefore, delinquency tends to be unrestrained by any kind of control (p. 658).

Delinquent subcultures arise within gangs since particular forms of illegal behavior are seen as necessary for the performance of the dominant roles necessitated by the values of the neighborhood (Cloward and Ohlin 1960:7). Cloward and Ohlin (pp. 161–186) argue that the type of gang that emerges depends on the degree to which the conventional and illegitimate opportunity structures are integrated. *Criminal subcultures*, in which gang activity is oriented toward illegal means of securing income, are most likely in neighborhoods in which there are close bonds between different age levels and between criminal and conventional residents (p. 171). *Conflict subcultures*, in which the effective use of violence is the primary source of status in the group, tend to arise in neighborhoods with low levels of age integration and severe limitations on both conventional and criminal opportunities (p. 177). Finally, *retreatist subcultures*, in which the consumption of drugs is emphasized, are most likely among youth living in integrated areas who nevertheless have failed to succeed in either conventional or illegitimate activities, or among youth living in disorganized areas who are unable to successfully gain status through violent activities; in that respect Cloward and Ohlin (p. 179) refer to such gang members as double failures.

It is difficult to overstate the influence of these three theories on the development of criminological and gang theory; a recent book of James Short, Jr. (1990:148), for example, defines gangs in part as "carriers of subcultures." Not only have the predictions and assumptions of Miller, Cohen, Cloward, and Ohlin been subjected to numerable tests but they have also inspired several large-budget federal programs aimed at delinquency prevention. Klein (1971:31), for example, observes that the differential opportunity approach of Cloward and Ohlin "almost became a national policy" during the late 1960s and 1970s.

However, some important criticisms have been leveled at subcultural theories of gang behavior in general, and these three theories in particular, on both empirical and logical grounds. Several studies have failed to document the existence of a system of crime-related norms, values, and beliefs that is unique to the lower class. The most important series of research was that of Short and Fred Strodtbeck (1965), who designed their study of Chicago gangs in part as a specific test of these three theories. At a very basic level, they note that they could not find a single "full-blown" example of the criminal gangs discussed by Cloward and Ohlin, or more than one predominantly drug-using (i.e., retreatist) gang in all of Chicago despite "highly motivated" efforts to do so (p. 13); this failure led them to question the generalizability of such phenomena. Hagedorn (1988:99–100) has also been extremely critical of the Cloward and Ohlin typology, arguing that they failed to take into account the age-graded nature of gang behavior. For example, while he considers all the Milwaukee gangs included in his study to represent "fighting gangs" (which would be analogous to the conflict gang of Cloward and Ohlin), he notes that the fighting usually occurred when gang members were relatively young. As members mature, "their interests turned more to fundamental problems of survival." Such findings suggest that any particular gang may be characterized by a mixture of activities that are distributed among the members on the basis of age.

The failure to uncover the three types of gangs discussed by Cloward and Ohlin may be a function of the different localities in which these groups were situated, for we have noted the pronounced geographic variation in gang behavior. The more crucial test of the subcultural thesis is the identification of particular norms, values, and beliefs that may encourage and support delinquent behavior in lower-class neighborhoods. Such studies at best provide only partial support to these theories. Short and Strodtbeck (1965:47–76) conclude that while gang, lower-class, and middle-class boys differ in their tolerance of behaviors proscribed by the middle class, the endorsement of middle-class values is "uniformly high" (p. 76). Even gang members evaluated images associated with a middle-class life-style more positively than any subcultural images, especially those that are illegitimate (p. 59). Overall, although there was some evidence of distinctive lower-class attitudes, these were definitely secondary to conventional values and beliefs.

A similar lack of support for the subcultural thesis is found in the work of Sandra Ball-Rokeach (1973).

Some criminologists have resisted the rejection of subcultural explanations, noting that there is a negative association between class and involvement in violent crime (Magura 1975; Wolfgang and Ferracuti 1982). Such arguments illustrate the primary logical problem of this class of theories. Marvin Wolfgang and Franco Ferracuti (1982:101) argue that since values can be identified in terms of expected behavior, "conduct is an external manifestation of sharing in (sub)cultural values." According to Short and Strodtbeck (p. 75), Miller inferred the existence of his focal concerns on the basis of such observational data. However, Ball-Rokeach (1975:836) has countered that even if a perfect association is observed to exist between social class and crime, it is incumbent on subcultural theorists to demonstrate that the primary source of that relationship is the existence of norms and values among the lower clases that encourage the commission of crime.

One important alternative explanation is that the prevalence of illegal behavior in particular neighborhoods represents the effects of situational exigencies caused by the social structure and is not a reflection of a "semiautonomous subculture" (Erlanger 1979:235). Without a demonstration of the independent effects of a distinctive group of norms and values, subcultural theories become exercises in circular reassoning; that is, people in lower-class neighborhoods commit crime because of a subculture that encourges such behavior, and the proof of the existence of that subculture is that people in those neighborhoods commit crime (see Kornhauser 1978:209). We do not in any way mean to deny the existence of subcultures in our society; given the diversity that exists in the United States, only a fool might suggest otherwise. Rather, the central question is whether the values that are characteristic of particular social groups directly require or condone illegal behavior (see Erlanger 1979:235).

Although there is some contradictory evidence in the literature,[16] we do not feel that the bulk of contemporary research presents a convincing case for the existence of a unique crime-based subculture within lower-class neighborhoods that can explain either the patterns of gang behavior or the development of an ongoing criminal tradition within stable neighborhoods. In fact, some have questioned the existence of any distinctive subculture particular to the lower class. John Reed (1982:142–143) has outlined four implications of a subcultural theory of behavior that can be easily adapted for our purposes. First, many residents of lower-class communities will take gang behavior for granted because it is a type of activity that they find "natural." Second, gangs will not engage in illegal behavior in all circumstances but only in those where the culture permits or demands such behavior. Third, gang members are not marginal members of the community. Rather they have been well socialized into the culture of the neighborhood that gives rise to gangs. Finally, if gang behavior is part of a community's "cultural bedrock," it should be a common theme in local interactions.

There is some evidence that these criteria may be met in certain lower-class communities. For example, one of the informants quoted by Skogan (1990:25) states that a local gang has "the blessing and support of their parents and of their community." Yet Skogan's discussion implies that such support is not derived from a unique set of norms and values that encourge such behavior. Rather, this neighborhood was faced with the structural threat of racial invasion; the primary function of this gang was to "keep blacks out of the area." Based on her extensive review of the literature, Kornhauser (1978:210) has questioned the existence of any shared subculture that would unite such diverse elements of the lower class as poor whites, poor Southern blacks, poor ethnics, poor farmers, and poor slum dwellers in a distinctive set of social relationships other than "their indifference to or hatred of each other." We concur with her assessment.

It might be argued that classes are an inappropriate focus of subcultural theories. Rather, perhaps subcultural theories should focus on the development of criminal traditions within ethnic communities, whose residents clearly share a commmon heritage. Joan Moore (1978:52–53) has described how Chicano barrios are characterized by a wealth of shared cultural themes and long-standing family relationships. In this context, she argues that barrio gangs in Los Angeles are fully integrated parts of the community that provide the means of attaining culturally valued goals, such as the expression of maleness, competence, and "being in command." Similar observations have been made concerning Mexican-American gangs in other Chicano neighborhoods (Horowitz 1983; Virgil and Long 1990). It may therefore be possible that in such ethnic contexts, gangs serve as the carriers of neighborhood traditions and cultures.

Again, the question is the degree to which such ethnic subcultures encourage and support illegal activities. Wolfgang and Ferracuti (1982:160), for example, suggest that violent behavior is most likely in subcultures that emphasize masculinity, noting that "the adult male who does not defend his honor or his female companion will be socially emasculated . . . [and] . . . forced to move out of the territory, to find new friends, and make new alliances." Ball-Rokeach (1973:740–742) finds no support for the proposition that attitudes associated with "machismo" are significantly related to violent behavior. On the other hand, although residents of the Chicago neighborhood studied by Horowitz (1983:22–24) aspire to achieve the American Dream through traditional means, she also notes that these values coexist with a code of honor that emphasizes masculinity, the style of one's actions, and one's personal integrity. The failure to elicit respect from others is perceived as a "derogation of fundamental properties of self" (p. 23), and the subculture of the neighborhood demands a physical and sometimes violent response to such personal affronts.

Horowitz (1983:Chapter 2) extensively discusses the degree to which gang behavior is the outcome of a cultural system in which group members appraise and evalute their behavior and that of others in terms of shared congitive and moral categories. Not surprisingly, she concludes that culture determines the meaning of social relationships, thereby giving rise to the gang activity in the area (p. 21). Since she is dealing with only a single neighborhood, it is impossible to examine the relative effects of structure and culture on the formation and maintenance of gang behavior, and the question is moot for the 32nd Street area. However, Horowitz (p. 35) does document the very weak linkages between the neighborhood and Chicago institutions that could provide resources to the area: "The political actors ignore the area . . . the school board does little to improve its deteriorating schools, the local banks and food stores charge high prices for goods and services, and the city urban renewal plan designated many of its buildings for destruction and its residents for removal."

As we noted earlier, one of the reasons for the historical popularity of subcultural theories has been the apparent inability of systemic, social disorganization approaches to account for the presence of stable neighborhoods with ongoing traditions of gang behavior. We feel that this is due to the emphasis of Shaw and McKay on only the private level of systemic control, especially as reflected in family structures and dynamics. Because of this emphasis, the social disorganization model is at a loss when neighborhoods with fairly stable patterns of family relationships are shown to have relatively high levels of gang activity.

However, a consideration of the parochial and public dimensions of systemic control may account for such phenomena. Interestingly enough, this systemic solution is implicit to Albert Cohen's critique of social disorganization theory, in which he states that "the organization which exists may indeed not be adequate for the effective control of delinquency and for the solution of other social problems, but the qualities and defects of organization are not to be confused with the absence of organization" (p. 33). Likewise, William F. Whyte (1981:273) concluded that the primary problem of Cornerville was not its lack of organization, but its "failure to mesh with the structure of the society around it." Therefore, a systemic model of gang activity would argue in part that it is likely to arise in areas in which the networks of parochial and public control cannot effectively provide services to the neighborhood.[17] For example, Gerald Suttles (1972:225) suggests that gangs may result from the inadequacy of the police, courts, and public schools to take responsibility "for the protection of property and lives and for moral education." These "urban service functions" of street gangs have been discussed exrensively by Martin Sanchez Jankowski (1991:180–201), who documents gang activities ranging from the protection of elderly neighborhood residents to preventing the construction of an unwanted real estate development.

There is some limited evidence that the development of parochial and public networks of systemic control can significantly modify the nature of gang behavior in a neighborhood. Gary Schwartz (1987) has extensively discussed the differences in two working-class communities in Illinois. Cambridge, a suburban community, was characterized by a great deal of distrust between local institutions (which were dominated by middle-class professionals) and the blue-collar residents of the town. In addition, pronounced generational tensions existed. Within this setting, drug use and fighting were widespread and took on the "moral equivalence of sports" (p. 144), especially among a group of youths who were highly antagonistic to authority, hypersensitive to insults, and fairly extensively involved in serious delinquency. These features are highly reminiscent of those implicated by Cohen's notion of reaction formation.

However, Schwartz discusses a second working-class community, the Parsons Park neighborhood of Chicago. In addition to its economic characteristics, Parsons Park is a primarily ethnic area that is being threatened by racial invasion. Therefore, it would seem to be an especially fruitful area for the development of a criminal tradition carried by local gangs. However, Schwartz's description gives no indication of any such behavior. While the adolescents in the area certainly get involved in delinquent activities, Schwartz notes that there is little serious juvenile crime. What does exist in the community, however, is a strong integration of family, church, and police relational networks and a clear sense of political power. It is not necessary to develop a subcultural theory to account for the antagonism toward authority that exists in Cambridge. Rather the comparison between it and Parsons Park suggests that the answer lies in the relative effectiveness of the private, parochial, and public bases of systemic control in each area.

A limited degree of support for the systemic explanation of criminal traditions can also be derived from certain incidents mentioned in the more standard gang literature. David Dawley (1992) and R. Lincoln Keiser (1969:9) describe a period during which the Vice Lords, one of the "supergangs" of Chicago, received external funding to develop community self-help and black consciousness programs. Almost all of Keiser's informants reported that "gang fighting had completely stopped" during this time (p. 11). Similarly, Howard Erlanger (1979) reports that there was a decrease in gang violence in the barrios of East Los Angeles between 1969 and 1972 when many gang members became involved with a locally based political movement.

These dynamics characterize a relatively small number of gangs and are most likely short-lived (see Short and Strodtbeck 1974). Nevertheless, such findings suggest that subcultural theories are not required to account for the existence of gang behavior in stable communities.

Neighborhood Gangs and the Urban Economy

One of the most common criticisms that has been made of contemporary gang research is that the dominant theories were developed during a period of American history in which a large number of low-skilled industrial jobs were available to immigrant gang members. These jobs provided the first steps toward economic mobility and assimilation. As we have noted several times throughout this book, the occupational opportunity structure of the United States has changed dramatically in that many urban areas have been characterized by significant decreases in the number of manufacturing jobs that are available to relatively low-skilled workers. Therefore, the structural sources of mobility that Thrasher or Shaw and McKay assumed to exist have narrowed considerably. The three classic subcultural theories that we discussed in the preceding section were also developed well before these shifts in the political economy began to exert their effects on urban areas. Therefore, some aspects of these traditional theories are grounded in assumptions concerning the nature of urban dynamics that are no longer relevant (see Hagedorn 1988).

No significant reformulations of gang theory were forthcoming until relatively recently, when a small group of researchers began to propose an alternative explanation for the apparent existence of gang traditions in urban neighborhoods. As opposed to the subcultural dynamics that were emphasized by the researchers discussed in the previous section, this new formulation focuses on the important contextual effect of the urban economy on the nature of the neighborhood dynamics. Although this issue has been addressed in a growing number of studies (see Horowitz 1982; Virgil 1983, 1988; Huff 1989; Fagan 1990; Jankowski 1991), the theoretical implications have been developed most fully by Joan Moore (1978, 1985, 1988; Moore et al. 1983), John Hagedorn (1988), and Mercer Sullivan (1989).

These approaches draw heavily from William Wilson's concept of "the underclass" (see Wilson 1980, 1987), which refers to those extremely poor populations that have been abandoned in the inner city owing to the exodus of the middle class; in later work (1988) he defines the underclass as residents of census tracts in which the poverty rate exceeds 40 percent. Although it does not necessarily have a racial or ethnic connotation, the term has typically been used to refer to low-income blacks and Hispanics who have been especially affected by the changing economic structure of urban areas. This is understandable, since the Hidden Employment Index developed by the National Urban League[18] indicates that 22.8 percent of all African-American workers were unemployed during the first quarter of 1991, compared to 12.3 percent of white workers (1991:1). These problems are especially severe for teenagers seeking employment: the Hidden Employment Index suggests that 58.9 percent of such black teenagers are unemployed, compared to 30.4 percent of the whites (p. 5); similar patterns have been noted for Puerto

Ricans and Mexican-Americans (Moore 1978; Hagedorn 1988). Given the changing political economies of urban areas, Wilson (1980:171) suggests that such disparities primarily are due to the differential occupational effects of these dynamics and the concentration of blacks in those occupations which have been hardest hit by the elimination of many manufacturing jobs.[19]

While "the underclass" is a very popular and controversial notion, it has been very hard to define precisely (see Will 1991). The most common use of the term within criminology has been in reference to men and women who are permanently excluded from participation in the primary labor market of mainstream occupations (Moore 1978:27–29). As a result, members of the underclass are forced to rely on one or more economic alternatives, such as taking low-paying temporary or part-time jobs in the secondary labor market, temporarily living off relatives or spouse equivalents, using transfer payments such as AFDC or General Assistance, joining subsidized employment programs like CETA, or becoming involved in hustling or street crime (Moore 1988:7–8).

Mercer Sullivan (1989) has provided a rich description of the implications of such economic dynamics on the development of gang dynamics in the "La Barriada" neighborhood of Brooklyn during the late 1970s and early 1980s. Her description of this area has all the classical features usually associated with underclass communities: most of the residents were supported by transfer payments, jobs that paid salaries just above what could be obtained through welfare, or jobs that were never officially "on the books." Economically motivated delinquency was a common behavior of young teenagers in the neighborhood, and such activities usually fulfilled a need for a short-term cash flow and excitement. However, as these adolescents aged, more regular sources of income were necessary, partly in an effort to support their families as well as their own life-styles.

Many of these youths were skeptical of the relevance of education to their future in the labor market and left school prior to graduation to obtain work.[20] However, given their relative lack of employment credentials, the postions that were available tended to be unstable, with undesirable working conditions and no chance for advancement. Since a significant proportion of these jobs were never officially recorded, many of these youths did not qualify for unemployment compensation when a position was suddenly terminated (Sullivan 1989:60–64). As a result, many of the local youths became involved in a systematic series of thefts and other economically motivated illegal activities that were coordinated through membership in local gangs (p. 117).

The image of gangs found in Sullivan's work goes beyond the play groups that are bound through conflict with other groups. Rather, as Jankowski (1991:22) has argued, "the gang emerges as one organizational response . . . seeking to improve the competitive advantage of its members in obtaining an increase in material resources." While Jankowski (p. 120) notes that gangs focus their economic activities primarily in goods, services, and recreation, "the biggest money-maker and the one product nearly every gang tries to market is drugs."

Sullivan (1989:136) notes that some of the barrio youths that she studied greatly decreased their involvement in criminal activities after the age of eighteen because of the greater opportunities that existed for legitimate employment. Although this pattern is identical to that assumed to exist by Thrasher, there are several important differences. First, some youths had difficulty obtaining legitimate employment because of the criminal records they had compiled due to their gang involvement. Moore (1988:10–11) observes that these problems are especially severe when Chicano gang members have served time in prison or a juvenile correction facility because they often return to the neighborhoods unemployable and with a history of heroin use. Thus, the prison experience can accelerate the process of economic marginalization.

Second, those youths who were able to find legitimate employment sometimes supplemented their small salaries with the proceeds of criminal activity, which increased the likelihood of further involvement with the criminal justice system and eventual dismissal from the job. Finally, owing to problems of excess absenteeism, many of these legitimate jobs did not have a long duration. Such employment histories would necessitate further participation in gang activities for the sake of economic survival. As a result, contrary to the gangs discussed by Thrasher and many subsequent researchers, some residents have maintained their gang involvement well into early adulthood.[21] Hagedorn (1988:110) makes the important observation that this is also the case for the "Norton" gang studied by Whyte (1981), who resided in an Italian neighborhood during the Great Depression when economic opportunities were similarly limited.

Moore (1988:8) has argued that such economic processes can lead to the institutionalization of gang activities in underclass neighborhoods. While a signifiant proportion of gang members may mature out of such behavior (as also noted by Sullivan), some fraction is recruited into gangs during adolescence, and the process continues to reproduce itself.

It might appear that the underclass hypothesis represents an important competing alternative to the systemic social disorganization model. Like the subcultural theories that we have discussed, it is able to account for the presence of gangs in otherwise stable, low-income neighborhoods. However, the patterns that have been described in this literature are very similar to those found in Spregel's (1984) discussion of the "segmented" route to gang behavior, in which he emphasizes the economic nature of gang activity. Spergel (p. 203) noted that secondary institutions in such neighborhoods "have efficient links to the local population as well as to each other." However, these institutions are secondary to those "representing dominant city and local middle class interests." Moore (1978:21–26), for example, notes that the general absence of political "brokers" who can intercede between underclass Chicano communities and major institutional agencies (such as those connected with health education and welfare, criminal justice, education, and immigration) has left residents poorly equipped to deal with such institutions. Even programs that titularly have been designed to bring resources into the neighborhood can be perceived as self-serving and alien if there is not sufficent input and participation by local residents. As a result of these relatively weak linkages to the centers of power in a city, the potential ability of an economically marginal neighborhood to exercise effective public systemic control is very limited.

Some might argue that it is very difficult for any particular neighborhood to significantly affect the political economy of its metropolitan area to such a degree that meaningful occupational opportunities become available to residents desiring such employment. This is certainly true, for Suttles (1972) has noted that such large-scale efforts are not likely to be successful unless neighborhoods can develop alliances with other groups in the city with similar interests. While this is often extremely difficult for areas with weak bases of public control, it is not impossible. Horowitz (1983:43) notes that the 32nd Street neighborhood engaged in several indigenous collective efforts to affect the local job situation. Likewise, city governments have been lobbied to provide incentives (such as tax breaks) to encourage local employers to remain in an area or to encourage prospective employers to relocate to an area (as was the case in the recent negotiations of several cities with United Airline). The success of such efforts can in turn further strengthen the public basis of control in an area, increasing the likelihood of success in similar future efforts.

Therefore, we do not believe this body of research contradicts the assumptions of a systemic model in any way. Rather, it is simply a difference in emphasis: the traditional social disorganization research has emphasized the private level of systemic control, the "underclass" work has focused on the public level. A fully systemic model, with a consideration of the private, parochial, and public orders of control, can account for the processes described by each set of theories in a logically consistent manner.

Conclusions

The development of a systemic social disorganizations description and analysis of the relationships between neighborhood dynamics and gang crime must contend with some serious criticisms pertaining to its apparent inability to account for the persistence of gangs in neighborhoods that are relatively stable. This situation has been compounded by the tendency of many traditional disorganization theorists, such as Shaw and McKay and to some extent Thrasher, to resort to subcultural models without a consideration of the logical difficulties that such an integration of perspectives entails. In addition, given the period of history during which the disorganization approach was initially developed in criminology, some observers have argued that it is unable to deal with many of the contemporary dynamics that characterize urban areas, especially pertaining to changes in the local economy.

Our basic argument has been that the apparent failure of the social disorganization approach is due to its traditional emphasis on the private level of control. Some of the empirical patterns that have been observed in the literature are simply incompatible with such an approach with its emphasis on residential stability. However, those patterns are consistent with a broader systemic orientation that considers the simultaneous operation of three types of control: private, parochial, and public. The expansion of the social disorganization model to fully consider the implications of these other forms of control provides a more general framework that is able to account for the dynamics that have been observed in modern American neighborhoods without resorting to other theoretical explanations that are logically inconsistent.

6

Neighborhood-Based Responses to Crime: Policy Issues

With few and isolated exceptions, the rehabilitative efforts that have been reported so far have had no appreciable effect on recidivism (Martinson 1974:25).

It is difficult to study a social problem with sustained interest unless there is an underlying assumption that the effort will contribute to its solution
—(Kobrin, quoted in Laub 1983:95).

Although we have no hard evidence concerning the issue, it has always been our impression that criminologists suffer from a significantly higher rate of occupational "burnout" than found in most academic disciplines. There certainly are a number of extraordinary researchers and criminal justice personnel who have maintained a high level of productivity throughout their careers. On the other hand, there also are some noteworthy examples of people who suddenly seemed to vanish from agencies or the pages of major journals after a period of extensive visibility, or who shifted their attention to issues only marginally related to criminology.

In addition to the many structural factors that may be related to this disciplinary attrition, it also reflects the fact that criminology is an inherently frustrating field. Given the enormous complexity of the phenomena with which we deal, it is impossible to provide easy solutions for the prevention of criminal behavior or for the treatment of offenders. Therefore, despite the existence of numerous programs that have been developed for such purposes, the success of these efforts has appeared to be limited at best. For example, while a National Academy of Sciences (Sechrest et al. 1979:102) report acknowledges that one "cannot say with justified confidence that rehabilitation cannot be achieved," the panel members concluded that "the research literature provides no basis for positive recommendations about techniques to rehabilitate criminals." The harshest evaluation of crime reduction strategies is found in Samuel Walker's review (1985:221) in which he concludes that "the best criminological minds of our time do not have anything practical to offer.... the intellectual and programmatic bankruptcy is truly nonpartisan." This conclusion that "nothing works" has been accepted by many citizens and disillusioned criminologists as truth.

Not all criminologists share such sentiments (see Cullen and Gilbert 1982); even Robert Martinson (1979) has tempered the famous statement that he made concerning rehabilitation programs that is cited at the beginning of this chapter. As we will discuss in this chapter, crime prevention programs are notoriously hard to implement and evaluate, especially when they are instituted at the neighborhood level. Nevertheless, there is a growing recognition in the field that particular programs have had significant, albeit limited, success in the prevention of crime (see, for example, Andrews et al. 1990a, 1990b; Lab and Whitehead 1990).

Neighborhood-based programs fall into two general, but related, categories. The more traditional programs have attempted to maximize the ability of local communities to exert effective forms of systemic control over the potentially illegal behavior of their residents. Recently, the focus has shifted from an emphasis on the offender to an emphasis on opportunities for crime (see Lewis and Salem 1981). Therefore, newer programs have been designed to utilize community resources to minimize the risk of victimization and/or the fear of crime of local residents. We will examine each of these approaches in this chapter. Prior to that endeavor, it is useful to examine a number of dimensions that characterize all programs regardless of their orientation.

An Overview of Neighborhood Programs

Although embryonic theories of community-based crime prevention were partly responsible for the rise of the settlement house movement in the 1880s and are implicated in the discussions of the Pittsburgh Survey (Kellog 1909–1914) and the West Side Studies (Russell Sage Foundation 1914), the dominant theories of crime at the turn of the century in the United States assumed that illegal behavior was the result of pathological characteristics. For example, all children who appeared before the Cook County Juvenile Court at this time were required to undergo physical and psychological examinations under the direction of the psychiatrist William Healy (Bennett 1981:89). Although Healy had a primary role in the development of psychological theories of crime and delinquency, was an early proponent of the life history approach, and is considered by many to be a pioneer in the creation of individualized treatment programs, Ernest Burgess (1923:659) noted that "his appreciation of the role of social factors went little farther than common sense."

This orientation changed significantly with the growing prominence of the urban research directed by Park and Burgess in Chicago that emphasized the relationship between human behavior and urban dynamics. This focus began to be reflected in a growing number of social programs with community orientation, culminating in the famous Chicago Area Project of Clifford Shaw and Henry McKay in 1932, which will be discussed extensively later in this chapter. While community-based crime prevention programs have dominated crime control efforts since that time, especially since the establishment of the federally funded Community Anti-crime Program in 1977 (see DuBow and Emmons 1981), "community crime prevention" has evolved into a vague, generic term that is used to refer to a great variety of programs.

The Initiation of Crime Prevention Programs

There is a great deal of variation in how a crime prevention program comes into existence. The majority of programs are initiated primarily owing to the efforts of local citizens or organizations that already exist in a community. Stephanie Greenberg and her colleagues (1985:136) note that one of the basic strengths of such indigenous efforts is that crime prevention activities can be tailored to meet the specific needs of a particular neighborhood. However, they also point out that many neighborhoods may not have the strong organizational base that is necessary to successfully implement these programs. In addition, neighborhoods without extensive connections to the wider community (i.e., weak levels of public systemic control) may not know how to obtain external funding for their activities, technical assistance, or other resources that would increase the likelihood of success.

Crime prevention programs can also be initiated in a neighborhood by groups not specifically identified with that residential area, such as local and federal government agencies (law enforcement agencies in particular), nonprofit groups, or corporations (Greenberg et al. 1985:134). These exogenous efforts have the advantage of creating local organizations that by definition have strong affiliations with external institutions. However, such associations may face a significant degree of resentment from local residents who feel that the organizers have a limited understanding of the problems they face on a daily basis. Such hostility can lead to very low rates of local participation, or even the rejection of a proposed program (Podolefsky 1983). For example, when Clifford Shaw approached leaders of the Russell Square neighborhood to recommend that they establish a Chicago Area Project–affiliated organization in their community, they were initially very skeptical of the Project and felt that Shaw's proposal repesented an attempt to portray the local population in a negative light (Schlossman and Sedlak 1983a, 1983b).[1] Likewise, Anthony Sorrentino (in Laub 1983:238) recalls how the Hull House (a famous settlement house located in the Near West Side neighborhood of Chicago) was perceived by local residents as a "superimposed" institution "run by a group of trustees on the Gold Coast."

While there are examples of local programs that have purely indigenous or exogenous sources, this distinction should be viewed as representing two ends of a continuum, for most modern approaches to community crime prevention are characterized by a combination of indigenous and exogenous elements. Unfortunately, as noted by Greenberg et al. (1985:136), there is very little reliable evidence concerning the relative effectiveness of internally and externally initiated programs.

The Dominant Orientation of Crime Prevention Programs

Community programs also differ in terms of their general orientation toward the prevention and control of crime. Some efforts are primarily concerned with locality development (Rothman 1979) and enlist a wide range of residents and local institutions to identify the needs and goals of the community. Although crime prevention may be the primary activity of the program, this is not always the case; some organizations may feel that the most effective approach to crime is an indirect one (Greenberg et al. 1985:125). For example, Aaron Podolefsky and Fred DuBow (1981) note the existence of crime prevention programs that have emphasized a vigorous enforcement of local norms, a clear delineation of neighborhood boundaries, increased interaction among the residents, and the development of a stronger sense of community. Such "social problem" approaches assume that crime will necessarily decrease as a neighborhood's capacity for informal control increases through these efforts. Other programs have combined such activities with those more directly related to crime prevention. For example, the Crisis Intervention Services Project (CRISP; Spergel 1986; Spergel and Curry 1990), a gang control project in Chicago that was based on similar programs in Philadelphia and Los Angeles, combined direct crisis intervention and mediation with gangs on the street with the broader mobilization of local organizations and residents.

A second general approach to community-based crime prevention, which has been referred to as social planning (Rothman 1979), emphasizes the local application of programs designed by experts with technical skills and specialized knowledge.[2] Many of these programs attempt to reduce the opportunities for crime in a neighborhood through such programs as property identification, block watches, or housing surveys (Rosenbaum 1987).[3] Others involve attempts to develop favorable relations between the police and local residents and efforts to increase police effectiveness in a neighborhood. For example, Wesley Skogan (1990) has discussed experiments conducted in Newark and Houston in which police-designed programs were implemented and evaluated. Obviously, while community solidarity might be a desirable secondary outcome of such efforts, the direct control of crime is the primary consideration.

The final approach is social action (Rothman 1979) in which efforts are made to mobilize economically and politically disadvantaged groups so that resources are redistributed and institutional policies are changed. While the control of crime may be a desirable outcome, it is not the primary goal of such programs and, in fact, may have a relatively low priority (Skogan 1988). The rationale for social action activities is well exemplified in a statement made by Saul Alinsky (1946:82), one of the most noted advocates of such programs: "You don't, you dare not, come to a people who are unemployed, who don't know where their next meal is coming from, where children and themselves are in the gutter of despair—and offer them not food, not jobs, not security, but supervised recreation, handicraft classes, and character building." Skogan (1988:42–43) has referred to such community programs as insurgent, as opposed to the more traditional "preservationist" crime prevention approaches, which attempt to maintain or strengthen the common interests and values that are assumed to exist in a neighborhood. Perhaps the most famous example of an insurgent program was the Mobilization for Youth project in the Lower East Side of New York City, which changed fairly rapidly from its original locality development orientation to one that emphasized direct and sometimes violent confrontations with the major institutions of the city. We will examine this program in more detail in the next section.

As was the case with the indigenous/exogenous distinction, these three orientations should be considered to be ideal types, for many programs combine elements of all three. In fact, the systemic model that we have been developing in this book suggests that a focus on a single orientation would be at best an incomplete attempt at social control in a neighborhood. A pure locality development program, for example, might result in strong networks of private and parochial systemic control, but relatively weak or nonexistent networks of public control. Since the social planning approach emphasizes crime control strategies such as surveillance and target hardening, a neighborhood's capacity for private systemic control may increase. In addition, since many of these programs involve the intense participation in the daily life of a community by an agency concerned with crime control, the capacity for public control is increased, at least for the duration of the program. However, there usually is very little emphasis on parochial forms of control. Finally, the institutional emphasis of social action programs may result in an increase in the levels of parochial (local institutions) and public (citywide institutions) systemic control. Yet, the development of the private level is not an intrinsic concern of such programs. A full community-based program of crime prevention would entail a concern with all three levels simultaneously. While such programs are rare, we will discuss a few notable exceptions later in this chapter.

Factors Associated with the Success of Crime Prevention Programs

There are two primary reasons why a community-based crime prevention program might fail. First, the crime-related activities might be grounded in an invalid theory of crime causation and/or community organization. This seems to be the most common conclusion that is reached when a program fails to attain its stated goal and is reflected in the statement of Samuel Walker concerning intellectual bankruptcy that we cited at the beginning of this chapter. When such opinions are expressed in the context of the conservative political climate that has been growing since the 1970s in the United States (see Andrews et al. 1990a), it is understandable that some people have rejected the relevance of sociological theories of crime causation and adopted alternative treatment strategies that are grounded in individual-level theories of biological or psychological pathology.

The ineffectiveness of some prevention and treatment programs can certainly be due to an erroneous theoretical foundation. However, before such theories are discounted as completely irrelevant, there must be a serious consideration of the degree to which the implementation of those programs deviated from that foundation. A classic description of the difference between the "ideal" program and its actual operationalization is provided by Paul Lerman (1975) in his evaluation of the California Community Treatment Program for juvenile offenders that was in opeation during the 1960s. As he notes (p. 54), it is impossible to evaluate the potential effectiveness of that program since in practice it departed significantly from its original design. More recently, Skogan (1990:146–149) has discussed a similar problem with the actual operation of the Minneapolis Community Crime Prevention Experiment (see Pate et al. 1987). In theory, this prevention program was designed to help local community leaders create a series of block clubs; each of these clubs was to have a strong crime prevention orientation. However, despite the great deal of effort that was exerted by members of the project staff, they did not get much response in about 30 percent of the blocks in the treatment area, and only 17 percent of the blocks fully met the program's organizational standards. The Minneapolis experiment did not appear to have any significant effect on neighborhood processes or problems (Skogan 1990:149). However, the problems that the staff faced in implementing the project make it impossible to determine whether this failure was due to a faulty theory or to its implementation.

In addition to the difficulties that are encountered in the design and implementation of community crime prevention programs, it has been notoriously difficult to evaluate their effectiveness reliably. Greenberg et al. (1985:120–122) have identified seven factors that prevent many evaluations from reaching firm conclusions about the effectiveness of neighborhood programs:

1. The frequency with which the program actually engaged in particular activities is rarely noted.
2. Many evaluations have focused on the ability of the program to successfully develop its crime prevention approach rather than evaluating the effectiveness of that approach as a crime reduction technique.
3. The programs tend to be evaluated over a relatively short period of time, generally through a comparison of crime rates one year before and one year after the implemention of the program.
4. Many evaluations fail to compare the changes that occur in a partiuclar neighborhood to those that characterize the similar communities without the program during the same period of time.
5. Since it is impossible to control all the events that may affect the crime rate in an area, special care must be taken to consider alternative, nonprogrammatic explanations for changes that are observed.
6. There are problems with measuring the outcome. For example, the apparent level of crime in a neighborhood may change simply because reporting behavior has changed or because the program emphasizes diversion from the kind of institutional treatment that would result in a recorded arrest.
7. Many programs are composed of a large number of concurrent related activities. Therefore, it is generally impossible to assess the effectiveness of any particular program component.

Arthur Lurigio and Dennis Rosenbaum (1986) also observe that many evaluations do not even utilize basic statistical tests of significance. Therefore, conclusions about program effectiveness usually must be made very tentatively.

However, two characteristics of successful community programs have been consistently noted. First, crime prevention programs have the greatest likelihood of success if they are integrated into the activities of more general, multi-issue neighborhood organizations. Skogan (1988:48–49) notes that crime is a "no win" issue in which it is difficult to document positive achievements and to maintain significant local participation over an extended period of time. Therefore, few organizations would care to base their chances of survival as a viable part of the community on this single issue. In addition, DuBow and Emmons note (1981:173) that few participants in crime prevention activities join an organization specifically for that purpose. Rather, they conclude that "the ability of a group to mobilize residents around crime issues may be a function of their success . . . at generating participation around the range of non-crime issues which already command the group's greatest attention." This phenomenon was recognized fairly early in the history of community-based crime prevention programs. For example, Shaw viewed the development of organized community recreation programs as a requisite first step in increasing adult involvement in the social activities of local youths. In turn, this involvement was expected to develop a sense of responsibility for those youths and, eventually, participation in programs oriented toward delinquency per se (Schlossman and Sedlak 1983a, 1983b).

The incorporation of crime prevention programs into the agenda of existing local organizations means that such activities can benefit from the strengths of the internal structures that have evolved over time. For example, Skogan (1988:55) notes that the first inclination of newly established organizations is to get some kind of exogenous assistance. Usually this is reflected in efforts to bring additional police resources into the community. Since such resources typically cannot be provided to individual neighborhoods on a long-term basis, these organizations must identify other agencies, institutions, or corporations that would be amenable to such requests. A local organization with an established reputation is more likely to be aware of such sources of assistance and may already have established relationships with these groups. Therefore, a structure amenable for the solicitation of external support is already in place. In addition, established organizations have had the time to develop effective leadership and to be characterized by paid staff, professional management, systems of financial accounting, and a recognized set of formal powers, all of which have been associated with successful crime prevention programs (Greenberg et al. 1985:156–157).

The second general finding is that community programs have been least successful in those low-income, heterogenous, high-crime areas that would ideally benefit most from their presence (DuBow and Emmons 1981; Skogan 1988). Rather, moderately cohesive, homogeneous neighborhoods, in which people can be mobilized on the basis of their interpersonal ties and which have at least a modicum of ties to resources outside of the communty, are most likely to benefit from such programs (DuBow and Emmons 1981:175). Not only do neighborhoods with weak or nonexistent sets of private, parochial, and public networks have high rates of program failure, but such efforts may actually increase the level of conflict within the community. For example, Skogan (1988:47) notes that in heterogenous areas characterized by hostility among the resident groups, citizen patrols may accentuate the divisions that exist. Again, given the general failure of the Chicago Area Project in such communities (Kobrin 1959), such findings confirm what became quickly apparent to the earliest community-based programs.

These considerations are crucial for the success of crime prevention activities, for it has been widely noted that the most effective programs are characterized by broad and representative local participation (Greenberg et al. 1985; Skogan 1988). However, even in the best of circumstances, rates of participation only range between 10 and 20 percent of the local population (Greenberg et al. 1985:139). As might be expected, some groups are more likely to participate than others, and many studies have reported higher rates of involvement in crime prevention activities by middle- to upper-middle-income persons, those who are married with children, home owners, the highly educated, those who are residentially stable, and the middle aged (see the review of Greenberg et al. 1985:139). While these patterns to some extent reflect the effect of such individual characteristics, they also are consistent with the relationship between community structure and program success that we have just noted. In fact, Skogan (1988:51) notes that characteristics of the neighborhood have significant effects on the likelihood of local participation in such events even after the role of these individual attributes has been taken into account.

There is one interesting deviation from this general pattern: blacks tend to have higher rates of participation than whites. Skogan (1988:51–53) argues that this is due to the more restricted opportunities for residential mobility that are available to black residents. Many white members of a community have three possible options concerning participation: join the program, do not join the program, and leave the community. Since black residents are less likely to be able to utilize the third option, they must either live with the local problems or attempt to do something about them.

Although they have rarely been considered as such, crime prevention programs easily can be considered to be neighborhood-based movements in which residents attempt to change some aspect of their immediate environment. When viewed from this perspective, the predictions of a systemic model of social control are very similar to those generated by the resource mobilization theory of social movements (McCarthy and Zald 1987). Before a movement can be successfully established, the structure of the neighborhood must be conducive to its establishment (Smelser 1962); potential members, a communication network, and local leaders are especially important (McAdam et al. 1988). In systemic terms, this means that a rudimentary network of private control must be established for the prevention program to have any type of effect. Therefore, the failure of most programs that have been established in low-income, heterogeneous neighborhoods is to be expected. The private level of control is also implicit to the resource mobilization notion of the "free rider" (Olson 1965), that is, someone who obtains the benefits of a program without active participation in the activities of the group. The free rider phenomenon is partly reflected in the generally low rates of participation that have been observed. Thus, this reflects a widespread phenomenon, and not one limited to crime prevention programs.

The resource mobilization literature can also provide other systemically related insights into the patterns that have been observed. John McCarthy and Mayer Zald (1987:16) note that the traditional social movement literature, which emphasized the role of grievances and beliefs concerning the redress of those grievances, has not been entirely successful in predicting the success of particular movements. The resource mobilization perspective emphasizes that a simple desire for change is not enough to guarantee success. Rather, a central consideration must be made of the amount and nature of the resources (personal, organizational, and financial) that are available to a group that desires change. For example, the perspective explicitly recognizes the necessity of effective linkages between social movements and other groups in its environment and emphasizes the crucial importance of involvement in the movement by individuals and organizations from outside the collectivity (see McCarthy and Zald 1987:16–19). The equivalence of these concerns to the systemic concepts of parochial and public networks should be obvious. The central relevance, then, of this body of literature is that prevention programs that do not concurrently utilize the resources inherent to their private, parochial, and public networks can expect a limited degree of success.

Overall, community crime prevention programs are not characterized by the sweeping effectiveness that would be desired. Sometimes this is due to the improper implementation of a program, sometimes because effective programs are not possible given the systemic structure of a neighborhood, and sometimes because programs have not exploited the full range of their systemic resources. In addition, owing to problems that have been noted concerning the evaluation of crime-directed programs, sometimes we simply don't know if a program worked or not. Greenberg et al. (1985:133) conclude that existing research "provides fairly strong evidence that local voluntary organizations can influence crime and fear of crime." While we agree with this assessment, the examples to be discussed in the following sections suggest that the success of such programs has been fairly limited.

Offender-Based Neighborhood Programs

Dan Lewis and Greta Salem (1981:405) have observed that prevention efforts during the first three-quarters of the twentieth century were characterized by an emphasis on the modification of the likelihood that people would participate in illegal activities. Alden Miller and Lloyd Ohlin (1985:11) note that those programs that had a neighborhood orientation shared the core assumption that the central resources for the prevention, control, and treatment of crime and delinquency were possessed by the residents of the local community and by the institutional services and agencies through which the community acted. This assumption is still at the heart of most contemporary community programs (see Rosenbaum 1986). To a great extent, these modern approaches represent modifications and adaptations of three classic prevention strategies.

The Settlement House Movement

The settlement house movement had its intellectual roots in the concept of the Social Gospel that was promoted by many influential Protestant theologians in Great Britain during the nineteenth century. Since all people were assumed to be united in an organic "human brotherhood of Christ" (Carson 1990:10), much of the suffering of impoverished urban populations could be alleviated by "bearing witness to a community of social interests that belied artificial economic distinctions" (Carton 1990:51).

The Social Gospel philosophy gradually began to take hold within the United States and eventually led to the creation of a number of settlement houses in economically deprived urban neighborhoods. This development was especially due to the efforts of Jane Addams, who had been strongly influenced by this philosophy (see Carson 1990:41) as well as by a series of newspaper articles concerning the London poor during a trip to Great Birtain in 1883.[4] During her stay she visited the Toynbee Hall settlement house, which became the model for her own Hull House, founded in Chicago in 1889 (Bennett 1981:111). While some settlement houses are still in operation (including Hull House), this movement thrived until shortly after World War I and then rapidly lost its momentum.

The primary mission of settlement workers in the United States was to facilitate the assimilation of the massive waves of immigrants who were entering the country at that time into the mainstream of economic and social life (Miller and Ohlin 1985:18). Therefore, houses first appeared in the cities that served as the primary areas of first location; by 1890, they had been established in Boston, New York, and Chicago (Carson 1990:10). To achieve the goal of assimilation, groups of "settlers" would move into houses in poor communities and attempt to become accepted by the local residents as neighbors and friends. After becoming established, the settlement house would offer organized, regularly scheduled activities for the residents (Carson 1990:52). In addition, the settlers would lobby civic institutions and agencies on behalf of the needs of their fellow neighbors, especially in the areas of housing, health, and employment (Bennett 1981:111).

Although one of the goals of these programs was the development of a sense of civic pride among the immigrant populations, the settlement house workers considered themselves to be mediators between competing social and economic interests. Therefore, the settlement house movement might legitimately be considered to represent primarily a series of social action programs. As a result, the activities of the houses were not especially directed at the prevention of crime and delinquency per se. Rather, as Robert Mennel (1973:155) points out, crime and delinquency were considered to be indirect outcomes of the poor living conditions that characterized these neighborhoods. It was assumed that once the environmental problems were solved, symptomatic social problems such as crime would eventually disappear on their own accord. However, there was some variation in this regard among settlement houses. One of the major projects of Jane Addams and the Hull House staff, for example, was the creation of a juvenile court in Chicago (Bennett 1981:111).[5]

Settlement houses also differed in the extent to which their religious roots were manifested in the community programs. Addams tried to avoid moral rhetoric and grounded most of her efforts in the findings of the urban sociologists at the University of Chicago, with whom she had developed a working relationship (Carson 1990:41; Miller and Ohlin 1985:18). However, religious themes commonly were expressed in the activities of other settlements. Steven Schlossman and Michael Sedlak (1983a, 1983b), for example, note that many residents of the primarily Catholic Russell Square neighborhood of Chicago resented the proselytizing that was said to occur at the Baptist-controlled Neighborhood House, which had been established in 1911. By 1932, Shaw concluded that this organization had only marginal significance in the daily life of the neighborhood (Schlossman and Sedlak 1983a:8).

The local attitudes that were expressed in regard to the Neighborhood House and its activities reflect a problem that was faced by all settlement houses to some degree. The settlement houses were created and staffed primarily by white, upper-middle-class, Protestant women (McBride 1975). Given the great social and economic gulf between these workers and the indigenous neighborhood residents, the relationship between the house and the community usually was tenuous (Finestone 1976:122). On the one hand, the staff could provide important connections to the local power structure that had not been available previously to the residents, and their efforts often helped foster a sense of pride in the local community (Sorrentino, quoted in Laub 1983:240). However, the settlers' assessments of the needs of the neighborhood were often developed within the framework of their own upper-middle-class values and priorities (Carson 1990:69). Sometimes they completely misjudged the nature of local norms, as reflected in a documented case in which a Christmas play was staged within a predominantly Jewish community (Carson 1990:64), or appeared to take a condescending approach to the neighborhood (McBride 1975:15) or some of its residents (Platt 1977:96).[6] Thus, Paul McBride has described the settlement house–community relationship as a "cultural cold war."

The settlement house movement began to disappear shortly after the end of World War I. Finestone (1976:121) notes that owing to financial problems, many houses sacrificed their organizational independence and became subsumed under larger, more centralized civic institutions. This trend was accompanied by a significant change in the self-identification of settlement house workers: self-proclaimed social reformers were gradually replaced by professional social workers. Therefore, the nature of this institution changed dramatically and those houses which are still in operation have a significantly different neighborhood orientation than that envisioned by Addams and her fellow visionaries. It is not at all clear what effect such programs had on local rates of crime and delinquency. However, it is indisuptable that all contemporary neighborhood crime prevention programs owe a deep intellectual debt to the community orientation of the settlement houses.

The Chicago Area Project

If a Hall of Fame existed for community crime prevention projects, the Chicago Area Project (CAP) would certainly be one of the charter inductees. Not only has it been one of the most widely discussed and controversial programs since its initial organization by Clifford Shaw (with the assistance of Henry McKay) in 1932, but it is also one of the most enduring and currently serves as an umbrella organization for over thirty local community associations in Chicago. While many of the components of the project may seem commonplace and mundane to contemporary community organizers, even Saul Alinsky, who sharply criticized Shaw for his failure to use confrontational strategies, has observed that the nature of the project's activities "were considered wildly radical then" (quoted in Sanders 1970:30).

CAP came into existence as a result of several factors. Most obviously, the systemic ecological work of Shaw and McKay had natural implications concerning the role of the community in the control of delinquent behavior. In addition, Shaw also had been engaged in an ongoing collection of life histories from delinquents (see Shaw 1930, 1931; Shaw et al. 1938). Although he never used this material to develop a systematic theory of crime and delinquency at the individual level (Bennett 1981:187), the life histories clearly illustrated the group processes that may give rise to certain delinquent events. In the context of his ecological work, Shaw felt that the life history data clearly showed that crime evolves as a "natural" phenomenon within certain community contexts. As such, Ernest Burgess and associates (1937; reprinted in Bogue 1974:85) described the techniques of CAP as analogous to the methods with which a physiologist deals with fatigue. That is, fatigue is almost an automatic outcome of certain physical contingencies. Just as a medical doctor attempts to eliminate fatigue by manipulating those factors that give rise to it, the CAP attempted to manipulate the characteristics of the community to which crime and delinquency was a natural response.

Finally, Shaw had become very disenchanted with the crime prevention efforts of settlement houses. Not only did he feel that such programs tended to isolate delinquent behavior from the full set of community dynamics that gave rise to such activity, but he also felt that the types of reforms pursued by settlement house workers would only have long-range, indirect effects on these dynamics. What he desired was a program that could have a more immediate impact on delinquency (Sorrentino 1959:41). In addition, he believed that crime prevention programs would only be meaningful to the majority of neighborhood residents if they exploited the strengths of the existing system of social relations to which the population had become accustomed (Kobrin 1959:29). Therefore, he rejected the tendency of settlement houses to arbitrarily impose middle-class values on impoverished urban communities (Pacyga 1989:162).

The combination of these three factors gave rise to the central prevention philosophy of Shaw and McKay and, in turn, the CAP: for delinquency prevention activities to be effective, they must first become the activities of the adults that constitute the natural social world of the juvenile (Kobrin 1959:22). Such activities would be designed to achieve three primary goals: bringing community adults into meaningful contact with local youths; exposing local residents to new scientific perspectives on child rearing, child welfare, and delinquency; and opening up channels of communication between local residents and institutional representatives who might provide useful resources to the neighborhood (Finestone 1976:127–128). Thus, despite the fact that the processes of systemic control were relatively undeveloped in the Shaw and McKay ecological model of neighborhood crime, the private (goals 1 and 2), parochial (goal 3), and public (goal 3) forms of control were not only recognized, but intrinsic to the operation of the CAP. Thus, it is the first fully systemic crime prevention program to have been developed in the United States.

The initial step in the establishment of a project in a local neighborhood involved the identification of the most powerful local institutions and the influential residents who were most familiar with the structure and history of the neighborhood (see the discussion of Kobrin 1959). This could often be a very slow process with many false starts. For example, in the Russell Square neighborhood, the site of one of the original three projects, Shaw initially believed that two local settlement houses offered the most promising opportunities for gaining access to residents of the area. However, he later discovered that most residents had an enormous distrust of these institutions, and he shifted his attention to the local Catholic church, which in fact was the most powerful indigenous organization in the area (see Schlossman and Sedlak 1983a, 1983b; Pacyga 1989). The second step was to convince these local actors that they and the community had a vested interest in the welfare of the neighborhood youth, especially in terms of their involvement in illegal behavior.

The third step was perhaps the key to the potential success of the project: the identification and recruitment of one of these community leaders to represent the CAP as an "indigenous worker" who would attempt to organize a broadly based group of influential community residents into a formal association whose central goals were focused on the welfare of the local youth, especially in the area of delinquency. Concerted efforts were made to develop strong affiliations between this association and the central local institutions of the neighborhood. Although the indigenous workers were made salaried employees of the Institute for Juvenile Research (which housed the CAP until 1957), their primary loyalties were expected to be to the local organization, and not to the Area Project.

Indigenous workers are commonly employed in many contemporary community prevention projects (see Spergel 1986). Yet, this was a revolutionary and highly controversial concept at the time the CAP was initiated. Kobrin (1959) notes that the use of local residents as community organizers has several significant benefits for neighborhood-based programs, for these workers already are familiar with the social life of the community and are likely to have ready access to the delinquent youths in the area. In addition, the central role of indigenous workers in the crime prevention projects demonstrated to the community that the CAP believed that the local residents were capable of self-regulation. This minimizes the amount of resentment that could be directed by residents toward "outsiders" who might be perceived as having a hidden agenda. For example, the residents of Russell Square initially were very suspicious of the goals of the CAP and circulated flyers that were strongly opposed to the program. The indigenous worker made arrangements for a public forum to be held at which such grievances could be aired. During this meeting, the highly respected workers defended the project in general and Shaw in particular and essentially gave a public blessing of the CAP. Schlossman and Sedlak (1983a:18–19) note that the forum played a central role in calming local fears and was essential to the project's survival.

However, the use of indigenous workers met with a storm of protest by professional social workers, who questioned the wisdom of endowing relatively untrained people with such responsibilities (see Bennett 1981:175). This hostility was further accentuated by the fact that some of the economic and political community leaders who served as resources for the project "did not always fit philistine specifications of respectability" (Kobrin 1959:footnote 4). Sometimes this criticism took the form of direct attacks against Shaw, as illustrated in a document written by a social worker from Northwestern University that questioned the poor Irish background of Shaw and accused him of being "rebellious and over-identifying with the poor and the sinners" (Sorrentino, quoted in Laub 1983:242–243). As Sorrentino notes, it took a long time for the conflict between the CAP and the social work establishment to be resolved.

Once the indigenous workers had developed a working organization, it was given a formal, structural autonomy by the CAP. A representative from CAP was assigned to each organization, and the local association could apply to the Board of Directors of the CAP for financial assistance, which was provided on a matching basis (Sorrentino 1959:44). Yet it was the responsibility of the organization itself to determine the nature and emphasis of its activities. For example, the neighborhood group could refuse to accept the CAP staff member who had been assigned to function as their executive (and could nominate their own executive, who was then hired by CAP) and could adopt any program policy that it wished. Even when this policy was considered unsound by the CAP staff and Borad of Directors, it was accepted nonetheless (Kobrin 1959).

The autonomous nature of the associations that were affiliated with CAP meant that there was no uniform CAP crime and delinquency prevention program. However, most of the community associations were involved in three primary activities (Kobrin 1959). Almost all associations had an organized recreational program, which was used to facilitate adult involvement in the everyday activities of the youths; it was assumed that this involvement would increase the sense of neighborhood responsibility for the behavior of their youths (Schlossman and Sedlak 1983a:10). Most local associations also instituted community improvement campaigns. Some of these efforts relied exclusively on motivating local residents to engage in activities that would lead to an increase in neighborhood pride, such as garbage removal, housing upkeep, and so forth. However, other programs pursued community development through the establishment of working relationships with institutions and agencies that either had a stake in community life or could infuse valuable resources into the area. For example, most communities appointed residents to act as liaisons between the neighborhood, the juvenile court, and law enforcement agencies. A system was often arranged in which these liaisons were notified if a local youth had been apprehended by the police; if at all possible, the liaisons attempted to have the youth released into their custody without the filing of a formal arrest report. Most neighborhood groups also worked closely with the local educational institutions and were often instrumental in developing P.T.A.'s in which the residents actively participated.

In addition, with the help of Shaw and the staff of CAP, some groups were able to obtain funding from local agencies. Many of these associations were primarily self-supporting; Pacyga (1989:176) estimates that two-thirds of the operating funds of the Russell Square Community Committee were derived from local sources. However, they also received assistance from CAP in successfully soliciting external support from the Chicago Community Trust, the Chicago Community Fund, and the Tribune charities. Organizations affiliated with CAP are still actively involved in the pursuit of local, state, and federal support (see Schlossman et al. 1984).

The recreational and community improvement programs formed the context for the efforts that were directed explicitly at the prevention of crime and delinquency. Again, the neighborhood groups were characterized by a wide variety of policies and programs. However, several approaches were fairly common. It was quite common for members of the association to regularly visit local youths who were institutionalized in reformatories or other correctional facilities. In addition, although the CAP is most widely known for its work with delinquents, most communities created programs designed to reintegrate adult parolees into the neighborhood. As was the case in many CAP-related activities, this was often done on an unstructured basis by making arrangements for housing and employment and sometimes providing some spending money. Shaw was very proud of the fact that the CAP programs spent a great deal of time dealing with such serious juvenile and adult offenders, for it reflected his optimism that community resources could be used to modify nearly all forms of illegal activity.

The most widely known prevention activity of most CAP affiliates was the creation of "detached worker" programs. These were young men assigned to the area by CAP (or, ideally, recruited from the neighborhood and put on the CAP payroll), who were expected to develop informal but intimate relationships with the youths in the area[7]; a special emphasis was placed on gaining access to those youths who were leaders in local street gangs. The detached worker was expected to spend as much time as possible with gang members during after-school and evening hours and to be alert to the presence of boys with serious emotional or mental problems, who might then be referred to the Institute of Juvenile Research for evaluation and treatment. However, the primary duties of the detached worker were "to provide additional structure for the boys' recreational activities, to be readily available for practical counseling and personal problem-solving, and to embody for emulation by local youth a model of 'conventional' moral and social values" (Schlossman and Sedlak, 1983a). Many widely distributed magazines were intrigued by the detached worker programs and popularized the colorful image of easygoing, sympathetic, young men who engaged in a great deal of "curbside counseling" (see Martin 1944).

Like the use of indigenous workers, many of the prevention programs evoked a great deal of controversy, not only among professional social workers, but also among the general public. One of the most sensitive issues concerned the occasional practice of placing serious juvenile and adult offenders into positions of responsibility within the associations. For example, an article in the popular *Harper's Magazine* (Martin 1944:509) noted that the president of a neighborhood center that was under the aegis of CAP had been paroled two years earlier after serving five years for armed robbery. Sentiments were especially aroused when Shaw offered state-financed employment to some of these parolees. When Shaw offered the afore-mentioned president an official position wtih CAP, thereby making him a state employee, he was criticized by professionals for considering an ex-convict with no academic credentials. These voices were joined by those of the general public who felt that such persons were being rewarded for their prior illegal activities.

In addition to criticisms that were leveled at certain programmatic elements of CAP, some have disagreed with its orientation on a more basic, theoretical level. Jon Snodgrass (1976:15), for example, has noted that the role of economic and political institutions in creating and maintaining the social conditions found in depressed urban areas was virtually ignored by Shaw. Rather, the CAP programs usually addressed these problems by developing the resources that already lay dormant in the neighborhood. When external resources were necessary, most organizations adopted Shaw's "pragmatic" (Snodgrass 1976:8) and nonconfrontational approach to institutional and corporate actors.[8] That is, they operated primarily within the existing systems of decision making and resource allocation. As Bennett (1981:175) has nicely described the Area Project, it had "a rhetoric of persuasion rather than of conflict." This conciliatory orientation eventually led to the resignation of Saul Alinsky from CAP, who went on to establish a set of community organizations grounded in more contentious operating principles.[9] Some observers have suggested that the orientation of CAP reflected the rural Indiana roots of Shaw and a desire to transform slum neighborhoods into urban versions of an idealized rural life (see, among others, Snodgrass 1976; Bennett 1981; Pacyga 1989). Others, such as Anthony Sorrentino (quoted in Laub 1983), have argued that the operational format of the CAP was designed explicitly to test the efficacy of the theory that Shaw and McKay had developed on the basis of their ecological and life history research. Regardless of the source of this orientation, the key question is whether it worked.

Unfortunately, the answer is not at all clear. It became obvious very quickly that the CAP approach did not work in those most disorganized neighborhoods in which all three bases of systemic control were extremely weak or vitually nonexistent (Sorrentino 1959:44; DuBow et al. 1979:70). For example, although it is often portrayed as a very impoverished community, the Russell Square neighborhood had a very rich institutional basis. More generally, it was difficult to document community changes that could be attributed directly to CAP activities. While Shaw initially was very outspoken concerning the effectiveness of CAP, his later statements were much more guarded (see Schlossman and Sedlak 1983a:110–114). Kobrin (1959) has highlighted two of the major difficulties in evaluating programs such as CAP. First, it is almost impossible to find neighborhoods that are identical in all respects except for the presence of a CAP program. Since an equivalent control group is not available, one can only attribute trends in crime to the effects of the CAP treatment with extreme caution. Second, without such control groups, it is generally impossible to isolate the impact of the CAP program on crime from the impact of urban processes that have affected the city as a whole. In addition, the evaluation efforts are completely confounded by the diversion activities engaged in by many CAP programs since a decrease in the reported crime rate could reflect the failure of illegal behavior to be recorded as well as an actual decrease in crime. Finally, given the great variation among CAP programs and the diversity of activities that are simultaneously sponsored within any single program, it is impossible to provide a global evaluation of the CAP approach or a more limited evaluation of a particular community organization.

Nevertheless, most of the extensive reviews of CAP have reached generally favorable conclusions concerning its effects. Harold Finestone (1976:139–140) has identified four criteria that increased the likelihood that a CAP program would be successful: the ability of the organization to raise funds, its degree of relative autonomy, the motivations of its members, and the relative integration of the members in the local community structure.

Throughout this section we have consistently used the past tense to refer to CAP since much of the material has pertained to events that occurred over thirty years ago. However, CAP is interesting for more than historical purposes since it is still very active in Chicago. Unfortunately, with the exception of the report of Schlossman et al. (1984), there have been few widely available discussions or evaluations of its current activities.[10] One trend noted by Schlossman et al. (1984) is that the CAP organizations now seem to be more likely to tailor programs so that they meet the guidelines of available federal funding. Therefore, some of the traditional emphases of CAP either have decreased in priority or have disappeared altogether. Schlossman et al. (p. 33), for example, note that this has been the case with activities aimed at the serious offender. Finestone (1976:137) has noted the tendency of some CAP-affiliated organizations to distance themselves from problems of crime and delinquency. In fact, during a meeting of representatives from all of the CAP programs in 1982, there was a long, heated, and at times violent discussion of whether CAP organizations should be concerned with crime and delinquency prevention *at all* (Bursik, personal observation).

The current and future status of the Chicago Area Project as a crime and delinquency prevention program is not clear at this point. However, especially through its use of indigenous and detached workers and its broadly systemic orientation to prevention, the influence of the Chicago Area Project on the design of subsequent community approaches has been indisputable.

The Mobilization for Youth Project

The Mobilization for Youth Project (MFY) was one of the first large-scale interventions into delinquency sponsored by the federal government and served as the model of delinquency prevention that was adopted by the President's Committee on Juvenile Delinquency and Youth Crime; certain components of the project eventually were incorporated into the War on Poverty during the 1960s (Miller and Ohlin 1985:20). While the Chicago Area Project is legendary among criminologists interested in the relationship between crime and neighborhood organization, its fame pales beside that of the MFY, which acquired a great deal of national notoriety for its activities. Despite the controversy that arose over the project, the theoretical basis and programmatic implications continue to have a major influence on the nature of contemporary community prevention approaches.

The MFY usually is considered to be grounded in the opportunity theory of Richard Cloward and Lloyd Ohlin (1960) that was discussed in Chapter 5, and they were involved intimately in its design and operation. Given its intellectual background, the MFY was designed under the assumption that juvenile delinquency could be decreased if the opportunities that were provided to youths through local neighborhood institutions could be brought into line with the aspirations of these youths; the primary targets were institutions concerned with housing, education, sanitation, employment, and law enforcement. These institutional changes could be accomplished if the adult residents of a neighborhood increased their degree of participation in local affairs and eventually moved into positions of institutional leadership, thereby holding "the reins of decision-making themselves" (Weissman 1969:23). This participation in local decision-making processes was expected to increase the identification of adults with the local community, which in turn would make them more likely to try to control the illegal activities of neighborhood youths. Therefore, delinquency was assumed to decrease with the increasing organization and integration of the community. Although there were no programs designed to directly deal with delinquent youth, a Legal Services Unit was established that served as an intermediary between the police, courts, and community (Helfgot 1981:77).

The history of the development of the MFY project is fascinating and has been recounted in detail by Joseph Helfgot (1981). Because of this history, Helfgot (pp. 16–18) argues that the MFY actually represented a hybrid of theoretical and prevention philosophies, including those that characterized the settlement house movement, the CAP, and Saul Alinsky's community organization efforts (see also Miller and Ohlin 1985:20). In fact, the MFY planners made at least one site visit to a CAP operation before the project was ever initiated (Helfgot 1981:18). To understand this development, it is necessary to consider the dynamics that gave rise to the project.[11]

The seeds of the MFY can be traced to 1957, when a group of representatives from agencies, institutions, and settlement houses that served the Lower East Side of Manhattan met at the Henry Street Settlement House. At that time, the Lower East Side was very unstable, had few indigenous institutions, had twice the unemployment rate of the city, and, most important, was the base of some very active and violent street gangs. A proposed program for local crime and gang control was eventually developed that was based on the organization of local agencies, institutions, and prominent neighborhood individuals into a Lower East Side Neighborhood Association. One of the central functions of this association would be to establish a series of neighborhood councils that would foster indigenous community controls and increase the identification of the residents with the community. In addition, teams of residents would be formed to intervene in gang activity, local volunteers would direct sports and recreation programs, and the association would provide services in the areas of mental hygiene, education, and vocational guidance. These programs would be developed with the active participation of public agencies, especially the police (Helfgot 1981:20). In its original formulation, the proposed activities of the MFY were very similar to those found in many of the CAP projects. In addition, as Helfgot points out, the initial proposal did not contain any references to opportunity structures.

Unfortunately, this group was not successful in obtaining external funding for the program owing to an undeveloped theoretical orientation and the lack of a research and evaluation component. They eventually approached the Columbia University School of Social Work, and Cloward and Ohlin agreed to assist them on the project. The proposal was submitted two more times before funding was arranged. According to Helfgot, the opportunity structure orientation of the MFY became an intrinsic part of the project in the final version of the proposal (see pp. 22–27). In addition, the School of Social Work began to exert a major influence over the direction of the project and eventually dominated its Board of Directors, making it more of an exogenous than indigenous program. The MFY was officially initiated in May 1962, sponsored by a three-year, $12.5 million grant jointly provided by the Ford Foundation, the city of New York, the National Institute of Mental Health, and the President's Commission on Juvenile Delinquency and Youth Crime.

However, it rapidly became apparent that few significant changes in the institutional provision of opportunities could be accomplished through the utilization of local resources, for the opportunity structures were controlled by forces outside of the community (Moynihan 1969:110; Helfgot 1981: 52–54). In particular, Helfgot notes that the organizers were especially disappointed in the general failure of the employment programs. Therefore, the project expanded its efforts beyond the boundaries of the Lower East Side and began to demand changes directly from the New York City government and its related municipal institutions. This change in focus was accompanied by an increasing use of confrontational strategies, such as rent strikes, school boycotts, and demands for personnel dismissals.[12] These strategies reflected a new philosophical orientation. For example, when the Legal Services Unit was originally organized, the legal system was seen as providing a service to the poor. However, the law was now seen as a "manifestation of power and the privileges and preferences of those who have the ear of power" (Helfgot 1981:78).

This shift in the priorities of the MFY had several important repercussions. Whereas it was originally founded as a delinquency prevention program, it came to focus its efforts primarily on eliminating the causes of poverty. As a result, the delinquency-related aspect of the project became a very low-priority item. In addition, through the ongoing conflicts with the municipal government of New York City, the MFY was becoming a threat to one of the institutions that had provided a significant amount of funding for the project. Finally, the protests that were sponsored by the MFY were highly visible ones that created a great deal of controversy among the general public. The very popular (and very conservative) *New York Daily News* accused the project of being a subversive organization and suggested that it had been infiltrated by Communists. Eventually, four separate investigations were made of MFY, one by the city, one by the state of New York, and two by the federal government. Although these investigations generally vindicated the MFY, the damage had been done and eventually most of their programs became administered through City Hall (Moynihan 1969:122). While many of its ideals continued to be reflected in the War on Poverty programs and especially the Model Cities Project (Miller and Ohlin 1985:21), the MFY ceased to exist as an independent, community-based delinquency prevention project.

Given the organizational history of the MFY, it understandably is difficult to evaluate what effects it may have had on the rate of crime and delinquency on the Lower East Side. Miller and Ohlin (p. 21) note that part of this difficulty arises from the inability to differentiate the impact of the project programs from the large social development programs that followed, or from the effects of the rapidly developing civil rights movement. In addition, the indigenous leadership resources of the Lower East Side were limited from the outset, and the technical capacity for orchestrating such a large-scale effort was lacking (Miller and Ohlin 1985:21). Overall, most summary descriptions of the program refer to it as a "noble failure" (Siegel 1986:213). However, like the CAP, its influence on the orientation of contemporary prevention programs is considerable.

Contemporary Offender-Based Programs

We have not provided extensive discussions of these three general programs in community crime prevention simply to illustrate the history of offender-based projects. Rather, given the extended periods in which they operated and the many careful evaluations that have been made of the assumptions and activities that characterized each approach, it is possible to provide a detailed sense of the strengths and limitations of each strategy. As we have noted, the programmatic components that were incorporated into the CAP and MFY in an effort to increase the levels of private, parochial, and public control are commonly found in contemporary prevention efforts: increasing the solidarity of a community, direct intervention, the use of indigenous and detached workers, the establishment of relationships with local institutions, and pragmatic and/or confrontational negotiations with municipal agencies. For example, the CRISP program described by Spergel (1986), the youth correctional program discussed by Miller and Ohlin (1985), and the Urban Initiatives Anti-Crime Program[13] (U.S. Department of Housing and Urban Development 1980) all have incorporated elements originally developed in the CAP or MFY. Although many fewer programs can trace their operational roots directly to the settlement house movement, the delinquency prevention program of the House of Umoja in Philadelphia has some striking similarities.

Nevertheless, while such broadly based efforts continue to be developed in the field of crime prevention, offender-based programs are not nearly as prevalent as they once were. Rather, especially since the mid-1970s, community crime prevention programs increasingly have been characterized by an emphasis on the potential victims of crime, and neighborhood-based activities are designed to reduce the opportunities for crime and, by assumption, lower the levels of fear felt by many urban residents. Part of this trend reflects shifts that have occurred in the priorities of funding agencies, part of it reflects a shift in theoretical emphasis due to the growth in popularity of routine activity and fear of crime research, and part of it reflects the premature rejection by some of the proposition that community resources can be utilized to produce significant modifications in the behavior of juvenile and adult offenders. Overall, despite the indications that such programs can have at least a limited effectiveness, it must be concluded that offender-based programs have passed their peak of popularity.

Opportunity-Based Neighborhood Programs

Beginning in the early 1970s, the Law Enforcement Assistance Administration (LEAA) of the federal govenment began to fund a number of projects that explored the viability of community efforts to minimize the risk of victimization within residential neighborhoods. This became one of the primary emphases of LEAA, culminating in the $30 million Community Anti-crime Program (CACP) in 1977 (Rosenbaum 1986:12) and later the Urban Crime Prevention Program (UCPP; Lavrakas 1985).[14] Community organizations were considered to be the primary agencies through which effective crime prevention programs could be developed, and funds were awarded directly to citizen groups and not to state or local governments (Lavrakas 1985:95).

DuBow and Emmons (1981:171) have summarized the four basic assumptions of CACP (and, by extension, UCPP) as follows:

1. Neighborhood residents can be mobilized by community organizations to participate in collective crime prevention projects.
2. Involvement in these activities creates a stronger community because

people will take greater responsibility for their own protection and local problems, and interactions among neighbors will be increased, both formally, through the activities of the crime prevention projects, and informally, as a by-product of these activities.

3. A stronger sense of community and increased social interaction leads to more effective informal social control.

4. Aside from the direct effects of community crime prevention activities in reducing crime or the fear of crime, these activities may also reduce crime or the fear of crime by rebuilding local social control in the neighborhood.

Although this orientation is nearly identical to that which characterized the offender-based programs discussed in the preceding section, there are two crucial differences. First, the CACP-sponsored programs were designed around the community control of criminal opportunities, and not criminals. Second, these programs emphasized the private level of systemic control and the development of informal forms of social control. There is relatively little consideration of the parochial level of control, and the manipulation of the public level of control in a neighborhood is restricted primarily to the development of an effective collaborative relationship with the local police departments (Skogan 1988:40).

Skogan (1988:65) describes the CACP as an "administrative quagmire" since local organizations did not receive any assistance in applying for these funds or guidance in spending the money that was awarded; likewise, although he notes that the UCPP had a more orderly operation, Skogan cites the example of a Chicago organization that spent an entire year preparing a proposal and lobbying for its approval (originally discussed in Lewis et al. 1985). Nevertheless, a large number of projects were funded by CACP and UCPP, as well as by similar programs instituted by the Ford Foundation. A number of other neighborhood organizations have sponsored related activities without any externally generated financial assistance. These programs generally have engaged in two distinct, but related, crime prevention strategies.[15]

Increasing a Neighborhood's Capacity for Surveillance

To many residents of urban neighborhoods, the police may represent a "thin blue line," that is, the final line of defense between law-abiding citizens and offenders. As George Kelling and James Stewart observe (1989), not only is this a gross simplification and distortion of the order maintenance functions of police departments, but it also is a relatively new public orientation to law enforcement. In the early part of the nineteenth century, individuals and neighborhoods had the primary responsibility for enforcing the criminal law by reporting local violations to the town marshal (Lane 1980). Police departments as we know them began to evolve when the dramatic growth in population that accompanied the urbanization of the United States made such a system unwieldy and prone to abuse; the first formal police department was not established in the United States until 1837 in Boston.

Citizen involvement in law enforcement has never completely disappeared, for an unpublished report by the Kansas City Police Department in 1977 indicated that citizen volunteers were responsible for 20 percent of the arrests that were made immediately after the commission of a part I offense (cited in Sherman 1986:348–349). Yet many communities increasingly have resisted the primary delegation of law enforcement activities to strangers and institutions that seem remote and unaccountable to the neighborhood (Kelling 1987:100). Therefore, with the encouragement and cooperation of local law enforcement agencies, some areas have developed programs in which residents act as the "eyes and ears" of the local police (Rosenbaum 1987:106). These neighborhood watch programs have been called the "centerpiece of community crime prevention during the 1980's" (Garafalo and McLeed 1989:326). The primary activity of these programs involves increasing the degree to which residents engage in the local surveillance of the neighborhood in an attempt to detect signs of criminal activity; upon such detection, the police are notified (Garafalo and McLeed 1989:327–329). This is usually coordinated through meetings of the community association during which information is shared about local crime problems, crime prevention tips are exchanged, and surveillance plans are developed (Rosenbaum 1987:104). These neighborhood/block watch programs are designed not only to increase the direct role of local residents in crime control activities, but also to foster the sense that they are "recapturing" the community from criminal influences.

The decentralization of law enforcement responsibilities has a very intuitive appeal. However, in practice, these programs face some very serious obstacles. Rosenbaum (1987) has noted that it is often difficult to involve a broad and representative cross-section of the community in these activities. This problem is especially acute in areas that are racially or economically heterogeneous since there may be a significant lack of consensus among local residents concerning the nature of the crime problem and the tactics that might be most effective (p. 117). In fact, Skogan (1988) observes that citizen surveillance patrols can often increase the tension that exists among mutually hostile groups in the neighborhood, and there is some evidence that the racial future of the neighborhood has emerged as a central topic of discussion at watch meetings (Lewis et al. 1985).

In addition, as we noted in Chapter 4, participation in these activities may actually increase the fear of local residents. It is hard to provide a definitive statement concerning this phenomenon, for the design of many evaluations has made it impossible to determine whether the positive correlation between fear and participation is due to selective participation or to the effects of the program itself. However, there is some evidence that participants in the Seattle Community Crime Prevention Program, which Rosenbaum describes as "exemplary" (1987:118), were characterized by a marginal increase in fear as a result of their involvement in the watch program.

Mixed findings have been produced concerning the effectiveness of watch programs. Of the three programs that generally are considered to have been most carefully designed and evaluated, that in Seattle is usually considered to have been successful, whereas those created in Chicago and Minneapolis are not (see Skogan 1990). Overall, Dennis Rosenbaum (1987:127) concludes that since there is little evidence that participation in watch meeting programs increases the likelihood of local interaction, surveillance, or interaction, these approaches generally must be considered to be ineffective.

Community Policing Programs

The second victim-based strategy differs significantly from the first in that the primary agent of control is the police department, not an indigenous community organization. We have included it in this section for two reasons. First, community policing programs represent attempts to foster strong relationships between local neighborhood groups and municipal institutions and the channeling of external crime control resources into the local community. However, although great efforts are made to encourage the participation of local residents, the impetus for these activities comes from outside the neighborhood (much like the projects undertaken by settlement houses). Second, many of these efforts include the creation of neighborhood/block watch programs, which are indigenously controlled despite their exogenous origins.

Lawrence Sherman (1986:364–365) reports that some departments considered the use of rudimentary forms of community policing after the publication of the President's Crime Commission report in 1967. However, they generally were not implemented owing to organizational constraints. The contemporary use of community policing strategies has been traced by Jerome Skolnick and David Bayley (1986:3) to the late 1970s and early 1980s, when it became increasingly clear that certain innovative approaches to neighborhood crime control (such as increasing the number of police in the area and using random motorized patrols) simply were not working. While saturating a community with police did result in a significant decrease in crime, this was a very expensive policy and the results were only temporary. The only technique that consistenly was shown to have an effect was the use of regular neighborhood foot patrols. However, while this police presence did reduce the level of fear among local residents, it did not appear to have any significant effect on the crime rate. Simply stated, something had to be done.

What emerged out of this general dissatisfaction was a reorientation of many departments from a reactive crime control philosophy to a positive crime prevention strategy in which the neighborhood was mobilized in support of the police efforts (Skolnick and Bayley:213). Skogan (1990:91) notes that this movement shifts the operational principles of police departments from a concern with crime fighting to a commitment to solving neighborhood problems. Thus, it envisioned crime control as a "co-production" of citizens and police (Lavrakas 1985:89); neighborhoods were no longer expected to be the passive recipients of law enforcement efforts and the police were expected to help the neighborhoods help themselves (Sparrow 1988; Skogan 1990).[16]

While a wide variety of community policing programs have been instituted on a permanent or temporary basis, all have been characterized by some combination of several elements (Skolnick and Bayley 1986:212–217). Perhaps the central attribute of all programs has been an emphasis on police-community reciprocity. This reciprocity represents the belief that the citizens have something to contribute to crime control. As a result, community organizations have actively been courted to work with the police to solve the crime-related problems of the neighborhood. In some cases, when indigenous associations were weak or virtually nonexistent, they have been activated by the police (Skolnick and Bayley:213). In addition, programs have been characterized to varying degrees by the decentralization of command posts into neighborhood locations and the reorientation of patrol officers to the neighborhood itself. Malcolm Sparrow (1988:6) observes that traditional patrol work has defined itself in terms of specific time periods; that is, officers are not responsible for what happens in the area when they are not on duty and can be dispatched anywhere in the city in the event of an emergency. The concept of community policing shifts this orientation to a geographic framework in which officers are expected to immerse themselves into the daily life and activities of the area. Finally, the most controversial aspects of some community policing programs have involved "civilianization" (Skolnick and Bayley 1986:217), in which community residents are formally given the responsibility for certain tasks that traditionally were handled by patrol officers.

The possible variations that can evolve in community policing programs are nicely illustrated by Skogan's (1990:93–124) discussion of the experimental programs that were instituted in Houston and Newark. The Houston police instituted small storefront "Police-Community Stations," created outreach programs accompanied by a series of community meetings, regularly patrolled problem areas, worked with school officials, instituted identification programs, distributed newsletters, opened a "safe house" for children, started a neighborhood beautification committee, and invited local representatives to ride with officers on patrol. Many of these activities were initiated and/or developed by a Community Organizing Response Team, which attempted to create a community organization where none had previously existed. In addition, a Citizen Contact Patrol program was started in which police officers would go from door to door and inquire about neighborhood problems about which they should be aware.

Some aspects of the Newark community policing program were very similar to those created in Houston: community police stations, a citizen contact patrol, a neighborhood newsletter, and a neighborhood clean-up program. The primary difference involved an intensive enforcement program, which focused on eliminating signs of social and physical disorder from the neighborhood. These efforts included street sweeps to reduce loitering and disruptive behavior, drug sales, purse snatching, and harassment, foot patrols, radar checks to enforce traffic regulations, bus checks to enforce local ordinances, and roadblocks to control automobile-related problems such as stolen cars and driving under the influence.[17]

The Houston and Newark programs had mixed levels of success. Skogan notes (1990:105–107) that the Houston effort seemed to be effective in lowering perceptions of physical and social disorder and in producing increases in local satisfaction with the neighborhood and police services. However, there was evidence that not all people in the community benefited equally from the program; the positive benefits were most noticeable for whites and homeowners. In Newark, on the other hand, the program did not appear to have differential effects in the community. The program generally led to an increased satisfaction with the neighborhood and police performance and a decline in signs of physical and social disorder and the fear of crime. The intensive enforcement component of the program, however, was only associated with a decrease in perceived social disorder and was actually associated with an increase in perceived physical disorder. Unfortunately, neither the Houston nor the Newark neighborhoods experienced significant decreases in victimization, and Sherman (1986:376) reports that the Newark area actually suffered an increase in the prevalence of household victimizations.

Overall, the use of community policing strategies has been shown to have positive effects on the quality of neighborhood life, especially in terms of perceived disorder and the fear of crime. However, the few evaluations that are available suggest that they are no more effective than traditional methods in the control of crime.

Policy Implications

In this chapter we have questioned the prevalent assumption that "nothing works" and have presented evidence that certain features of offender- and opportunity-based crime prevention programs have been shown to be partly effective in the control of crime. Unfortunately, while it is extremely difficult to gauge precisely the magnitude of this effectiveness, none of the programs we have discussed have resulted in an extraordinary degree of success. More critically, no program has been designed that has any degree of success in those neighborhoods with the highest rates of crime. Thus, it is understandable why the general public and many criminologists are extremely pessimistic about the future of crime prevention and control.

However, several observations may be made in response to this position. Most basically, it has been shown that certain approaches have had at least *some* discernible effect. Yet, while the theoretical justifications for most crime prevention activities are fairly persuasive, very little consideration has been given to the interaction of the various program components. Rather, it may be the case that the effects of some individual activities counterbalance each other, leading to a general conclusion that the entire program is a failure. For example, Donald Black and M. P. Baumgartner (1980:195) argue that residents will feel less of a need to draw on their own indigenous resources to control crime as police activities increase in frequency and visibility. As a result, the level of informal control that is exerted in a neighborhood may decrease when formal processes of control are perceived as increasing in effectiveness. The outcome of these two contradictory sets of dynamics could result in a net change of zero, leading to the conclusion that the entire program was a failure. Therefore, while a great deal of attention has been given to the implications of particular types of prevention strategies, it is necessary to consider the implications of particular sets of activities that often are concurrently utilized. The theoretical implications of such combinations are much less clear.

There may be a more general policy-based reason for the apparently limited effectiveness of most community-based crime prevention programs. Traditional programs, such as the Chicago Area Project and the original form of the Mobilization For Youth, strongly emphasized the role of informal associations and networks within the local community (see the discussion of Glazer 1988:140). In the words of Albert Hunter and Suzanne Staggenberg (1988:253), such networks are a "structural precondition" for the successful mobilization of a community to combat crime. The work of Anthony Oberschall (1973:125) also suggests that the likelihood of a successful program is highest when an association is able to recruit blocs of people who are already organized. Yet, as we have noted, those communities most in need of effective crime control programs often are those characterized by a very segmented set of networks that may be difficult to unite in a collective effort. Thus, the programs of the Chicago Area Project have tended to be very ineffective in highly unstable, economically marginal neighborhoods. Therefore, if it is nearly impossible to develop the internal structure that Hunter and Staggenberg have deemed necessary to a successful community effort, how would it be possible to institute any kind of successful crime-related program in such neighborhoods?

The proposal that we offer is sure to generate a great deal of criticism and controversy, yet it has been shown to have at least some potential for effectiveness—the recruitment of gang members as core members of locally based crime prevention programs. Clifford Shaw long ago recognized that people who have been involved intimately in the criminal activities of a neighborhood are an invaluable source of knowledge concerning the extent of crime, its distribution throughout the neighborhood, and those who are most involved. Since such individuals generally are embedded in relatively broad networks of association, the networks that are needed to form the foundation of a successful local organization do not have to be artificially developed by sources based outside the neighborhood. In fact, as Shaw eventually decided, it might be extremely efficient to make such gang members paid officers of the association.

The existence of such gang-based networks, of course, is not a sufficient condition for the development of effective programs, for the members must be made to identify with the crime-control sentiments that may underlie the program (Hunter and Staggenberg 1988:253). It certainly would be Pollyanna-ish to assume that such a task could be accomplished easily. Nevertheless, the notion of "turf" that has been emphasized since the work of Thrasher suggests that there is at least an incipient degree of identification with the neighborhood among gang members. In fact, historical precedents exist that indicate that such identification can evolve into a source of strength for the local community. For example, in the beginning of the twentieth century, the Bridgeport neighborhood of Chicago was a low-income, poorly educated Irish community characterized by several gangs. The most notable of these was the Hamburgs, who were implicated as key violent participants in the 1919 race riots of Chicago. Nevertheless, the existent structure of the Hamburgs made the gang very attractive to local politicians, who recruited members of the group to organize the neighborhood. The utilization of the Hamburgs as a local political organization had great benefits for the Bridgeport neighborhood in that many political appointments were made to local residents. Several decades later, three consecutive mayors of Chicago came from the Bridgeport neighborhood, including perhaps the most famous past member of the Hamburgs—Richard J. Daley, Sr. (see O'Connor 1975; Royko 1971).

A second powerful illustration can be drawn from the history of Conservative Vice Lords of Chicago, who have been noted earlier in this book.

During the latter part of the 1960s, several of the older members of the Lords generally became disenchanted with the daily existence in which they were enmeshed and attempted to rechannel the gang's activities into programs aimed at providing desperately needed services to the residents of their West Side Chicago neighborhood (see the discussion of Dawley 1992). They were successful in soliciting external resources to support these efforts and, at least according to two accounts of this period in their history (Keiser 1969; Dawley 1992), there does appear to have been at least a temporary decrease in the level of crime in the area. Unfortunately, when these funds disappeared, the community returned to its original status.

Therefore, gangs appear to be an optimal (and perhaps the only) foundation for the development of broad local community networks capable of exercising systemic control in highly unstable and/or historically impoverished neighborhoods. In fact, efforts toward gang-initiated community organization appear to be developing in Los Angeles as this book is being written. Although the Bloods and the Crips, easily the two most notorious gangs of contemporary Los Angeles, have been bitter enemies for over two decades (Terry 1992), the riot that was triggered by the acquittal of the four police officers accused of beating Rodney King has led to an attempt to form a community-oriented alliance between the rival groups.[18] While some observers argue that the truce represents a sincere effort to stop the violence in South-Central Los Angeles, others are skeptical and counter that it is simply an effort to join forces against the police (see Terry 1992). Regardless of the motivations that led to the negotiation of the pact, the current agreement is extremely fragile given the long history of conflict between the two groups and the failure of a series of past attempted truces. For example, gang killings continued at the same time that picnics were being held to encourage cooperation among members of the Bloods and Crips (Associated Press 1992).

Nevertheless, leaders of the two gangs jointly issued a plan that proposes a set of programs designed to meet the needs of their community (Cockburn 1992). The goals of this project are very similar to those developed by the Vice Lords in the late 1960s and include the gutting of burned and abandoned structures, the construction of career counseling centers and recreation areas, the beautification of the neighborhood, the rehabilitation of existing educational facilities, upgraded curricula, tutoring programs, increased teacher salaries, new health care facilities, the replacement of welfare programs by state-sponsored employment, programs for economic development, and an increased role of residents in law enforcement activities. The community orientation of this plan is reflected forcefully in the key slogan of the proposal: "Give Us the Hammer and the Nails and We Will Rebuild the City" (Cockburn 1992).

This gang coalition may not be able to generate enough resources from within the community to maintain the organizational structure needed to institute and maintain the proposed programs (Hunter and Staggenberg 1988:255). Therefore, the critical key to the programmatic success of the Bloods/Crips porposal will be the ability to obtain resources from outside the local community to fund these efforts. In addition, the proposal clearly recognizes that there are certain "fundamental" causes of neighborhood decline that are out of the direct control of isolated local associations (Skogan 1990:172). As Lewis and Salem (1986:124) point out, the development of associational ties to organizations outside of the community who share the same self-interests often can foster the ability of groups to take collective action against the type of problems noted in the proposal. This ability is illustrated in Jenkins and Perrow's (1977) comparison of the efforts of the National Farm Labor Union and the United Farm Workers to obtain union contracts for farm workers. Although both groups utilized very similar tactics, only the United Farm Workers were able to mobilize resources from other groups to which they were weakly allied. As a result, while the efforts of the United Farm Workers were successful, those of the National Farm Labor Union were not. The success of the current Bloods/Crips coalition will depend on their ability to mobilize such support.

The development of ties to established community associations in other neighborhoods that share an interest in crime control can be crucial to the success of a locally based crime control program, for these groups may already have established relationships with local politicians and business leaders who may be able to provide the resources necessary to an association's success (see Mayer 1983:155–156). Thus, networks with other community groups may foster the ability of an association to development public control in a neighborhood. In addition, collective action by a number of neighborhood groups with the same interest in crime control may mobilize a sufficient amount of shared resources to pressure local government and industry to provide funds for desired programs. Finally, some organizations may not have the political experience or sophistication to successfully compete for resources from business and government. For example, as discussed in this chapter, one of the reasons for the demise of the MFY was its inability to successfully negotiate with institutions outside of the Lower East Side without threatening the existence of the entire program. Loosely connected coalitions of such groups can provide access to the necessary technical and political expertise that will increase the chances of successful negotiations with sources of funding. Again, the formal integration of gangs into such programs may foster the development of such networks since many such groups have weak alliances with similar groups dispersed throughout the city.

In sum, we are proposing that for local communities to combat their crime problems successfully, programs must be developed that strengthen the ability of the neighborhood to develop strong levels of private, parochial, and public control. As has been assumed in most traditional programs, the cornerstone of these activities is the development of strong networks of association among the residents of a community, and between those residents and existent local institutions. The major problem that has arisen in the past is that the most successful programs have been able to build on existing networks. Such networks exist in the form of gangs even in those neighborhoods in which the likelihood of a successful program would seem to be the lowest. Although it seems counter to conventional wisdom, gangs may, in fact, provide the key to successful crime control in areas that have been assumed to be lost.

Second, it is absolutely necessary for neighborhood groups to develop the ability to access resources that are controlled outside of the neighborhood and channel them into the community. The most viable approach to the development of such relationships would be the establishment of ties to other community groups that have the same self-interests. The collective efforts of these groups would have a much higher likelihood of successfully gaining external resources than the efforts of any single association. On the basis of the systemic model that we have developed in this book, the failure of any crime-related program to focus simultaneously on the private, parochial, and public levels of systemic control in the neighborhood will heighten significantly the probability that such a program is eventually judged to be a failure.

Epilogue

Our intention in this book has not been the provision of a "definitive" test of the systemic model. Rather, we have simply attempted to show that a fairly simple, parsimonious approach could at least account for empirical patterns ranging from gang activity to the fear of crime that have appeared in the literature. The future development and refinement of this perspective faces a basic, critical problem. Throughout this book we have taken special care to point out the problems of measurement that have been endemic to this area of research. Measures of many of the key concepts implicated by a systemic model are not readily available to most researchers, and their collection would be extremely time-consuming and expensive. In fact, it literally may be impossible to incorporate an ideal systemic model into study designs that have the scope of the traditional social disorganization research. It sometimes has taken researchers several years to explore fully the systemic structures of a single neighborhood (such as the work of Suttles in the Addams area of Chicago). Even in a city with as few as fifty neighborhoods, the logistics of such a systemic study boggle the mind. Some researchers have made creative use of existing datasets to gain a sense of the systemic dynamics, as shown in Sampson's work with the British Crime Survey. Nevertheless, these studies often have been forced to rely on indicators that are extremely indirect and questionable reflections of the nature and dynamics of systemic control.

At one time, many criminologists lamented that the theoretical models which had been developed completely overwhelmed the capabilities of popular statistical technologies. That period has passed, and many criminologists routinely utilize sophisticated models that would leave researchers of even ten years ago shaking their heads. Unfortunately, the same problem continues to exist regarding the measurement of the key components of our theories. There is no doubt that significant advances have been made in the area of measurement theory. Nevertheless, the resolution of these measurement issues is the greatest single challenge faced by the contemporary criminological community.

Notes

Chapter 1

1. Throughout this book, the terms "neighborhood," "residential area," and "local community" will be used interchangeably.
2. Apparently, the stigma attached to the presence of pickup trucks has more than a regional basis, for similar legislation was also introduced in Flossmoor, Illinois.
3. The authors must admit a certain degree of sympathy for the antagonism that arose in the neighborhood when the traditional Italian sausages and *pierogis* of the area were being supplanted by blood orange salads with green mayonnaise dressing.
4. A large body of research has examined the relationship between an offender's place of residence, the victim's place of residence, and the location of the incident. See McIver (1981) or Costanzo et al. (1986) for reviews of this literature.
5. For important exceptions, see Hunter (1974) or Fischer (1982).
6. Hunter and Suttles also refer to nominal communities as "defended" communities. We have avoided this usage and reserve that term to refer to a particular phenomenon that will be discussed in Chapter 2.
7. An excellent review of these various perspectives is provided by Sheley (1991).
8. Throughout this book, "community" has been used synonymously with "neighborhood," that is, to represent a geographically bounded area of a city. However, if the notion of community is broadened to represent groups with social and/or symbolic, rather than spatial, boundaries, the dynamics discussed in this chapter might in fact be relevant to a much broader spectrum of criminal behavior. For example, Bursik (1988) has argued that if the focus of attention is shifted from the neighborhood to an organization that is assumed to be interested in self-regulation, then similar processes might apply to white-collar crime.

Chapter 2

1. James Bennett (1981) notes that in conversations conducted with McKay during the mid-1970s, McKay emphasized that the biggest achievement of the research

he conducted with Shaw was "opposing naturalist theories of delinquency and racial inferiority" (p. 320, footnote 37).

2. It should be noted that several studies have attributed much of this increase to changes in the way the police counted crime after the appointment of Orlando Wilson as Chief of Police in Chicago in 1961 (see Campbell 1969; Chamlin and Kennedy 1991).

3. Such widely used measures as median income or the average level of educational attainment are therefore compositional variables. The reliance on compositional measures is a common limitation of all sociological studies of group dynamics, not just those found within criminology.

4. The approach used by Gottfredson and Taylor is referred to as a covariance approach to contextual analysis. A sheaf coefficient is then used to reflect the overall effect of the neighborhood context.

5. Although this conversation may sound apocryphal, it did in fact occur. Out of the great respect that we otherwise have for this colleague, the source of the quote must remain anonymous.

6. These series pertain to the distribution of offenders, which represented their primary orientation. They also collected data for two additional, but decidedly secondary, series of data, pertaining to the number of cases representing male referrals to Juvenile Court between 1917 and 1923 and between 1900 and 1906 (as opposed to the number of males referred regardless of multiple referrals).

7. The image of Shaw, McKay, and their colleagues bending over a map, pens in hands, plotting the addresses of nearly 90,000 court referrals (and several times that number when all of their series are considered) is nearly inconceivable to those of us who have become used to address-mapping software. This model of patience and dedication should give pause to all researchers when complaining about slow turnaround times on their personal or mainframe computers.

8. City blocks have been used by Roncek (1987a, 1987b). However, his crime rates are not computed on the basis of the residential area of the offender as are the others discussed in this section.

9. It is important to note that such findings are mediated by the historical dynamics of the particular city under examination, and the ecological variables do not have simple, unidirectional effects on crime and delinquency. The implications of these findings will be addressed later in this chapter.

10. McKay notes in his afterword to the 1969 edition of *Juvenile Delinquency and Urban areas* that he and Shaw could also not foresee other important trends, such as the effects of open housing decisions and the greatly expanded role of urban planning programs.

11. Readers are sure to notice that we have not addressed one of the most important aspects of public control—the police services supplied to urban neighborhoods. This topic is addressed fully in Chapter 6, which discusses the efforts of neighborhood organizations in the control of crime.

Chapter 3

1. While very little is known concerning the selection of neighborhoods suitable for criminal behavior, there is a growing literature concerning how offenders

evaluate the characteristics of potential individual targets (see Bennett and Wright 1984a, 1984b; Bennett 1989; Carter and Hill 1979; Cromwell et al. 1991; Feeney, 1986, Rengert and Wasilchick 1985; Rengert 1989; Reppetto 1974; Walsh 1986).

2. In fact, Hindelang et al. (1978:241) define life-styles as routine daily activities.

3. The notion of goods and services is a very broad one in Hawley's model, for they range from the making of a certain tool and the distribution of manufactured products to the maintenance of law and order and the provision of education (p. 214).

4. Although work-related nondiscretionary time blocks provide the most obvious example of routine activities that may be conducive to crime, leisure activities also provide such opportunities to the extent that they are predictable. For example, many people go bowling or play poker at regular times of the week.

5. For example, Rengert and Wasilchick find that the daily activities of burglars were highly structured even among those who were unemployed.

6. Note that while social disorganization research is concerned with the addresses of offenders, routine activities research is concerned with the addresses of offenses.

7. For example, Garafalo notes (p. 82) that during the first half of the 1980s, the average number of rapes that were reported in the entire national survey was only 125.

8. Even if the NCS interviewed a sufficient number of respondents to make such a neighborhood-based analysis feasible, the procedures of the U.S. Bureau of the Census (which administers the NCS) prohibit the release of individual records for geographic areas with less than 250,000 residents.

9. The NCS asks a series of questions concerning criminal victimizations that occurred at the respondents' homes, near their homes, or near the homes of friends. However, the structure of the survey makes it impossible to determine the location of those that occurred in commercial places, in parking lots and garages, or in open areas/streets/public transportation.

10. It is important to note that while the geographic location of crimes reported to the police represents the place of occurrence, the recorded location of arrests is the residence of the offender. As we have argued, this second set of data is not appropriate for tests of the routine activities model, and the reader who reviews past studies (especially those of Schmid) must keep this distinction in mind.

11. While Sampson and Lauritsen (1990) also present a contextual model, their neighborhood variable is limited to the ecological proximity to crime.

12. This was also the case of the data used in Smith and Jarjoura (1988, 1989), which were drawn from an evaluation study of police services.

Chapter 4

1. The survey asked the respondents if there was anything about their neighborhoods that bothered them. If they responded yes, they were asked what it might be. The answers were open-ended and later categorized by coders. The respondents could provide as many responses as they wished, and each problem was recorded.

Of the 27,029 urban residents who noted a problem, 5,607 mentioned noise, 5,501 mentioned traffic, and only 3,968 mentioned crime.

2. Rachman (1990:3) considers fear to reflect a feeling of apprehension about tangible and predominantly realistic dangers, whereas anxiety is a reaction that is difficult to relate to tangible sources of stimulation.

3. Fear is considered to reflect an emotional response to possible violence crime and physical harm, whereas worry is a response to potential property crime.

4. The fear of crime measure used in the Hunter and Baumer study combined indicators of perceived risk and worry.

5. In Hartnagel's study, the fear of crime is assumed to affect neighborhood integration, rather than represent an outcome. The findings reported here represent the zero-order correlations presented in the paper. This work will be addressed more fully in a later section of this chapter.

6. We will reserve our discussion of the manner in which this behavior came to be classified as "gang-related" for the next chapter. At this point, we will only note that the classification in this situation was at best problematic.

7. Although we sympathize with the concern of the principal, it is hard to imagine a typical urban youth gang whose primary graffiti-based message is "Tax the Churches," as was the case in this incident.

8. Such a prediction can be traced directly to the arguments of Emile Durkheim concerning the functions of crime. These themes have been explored in detail by Conklin (1975) and Liska and Warner (1991).

Chapter 5

1. Those readers interested in a broader discussion of gang dynamics are urged to consult Spergel's (1990) review.

2. For example, our colleague John Cochran has referred to certain fraternities on the University of Oklahoma campus as "syndicates of rate," which has made him a very popular figure in the Letters to the Editor department of the school newspaper.

3. Interestingly, at least in that high school, this meant that the dominant gang continually revised its preferred style of dress to maintain a symbolic separation between it and the general school population.

4. If the Soviet Union had had five-year plans as effective as those attributed by many authorities to street gangs, it might have survived.

5. The careful reader will note that this problem is very similar to the compositional effect–group effect issue that was discussed in Chapter 2 in respect to neighborhoods.

6. A very similar argument concerning the role of gangs in adolescent development has been presented by Bloch and Neiderhoffer (1958).

7. Items included in the final scale reflected whether or not the group meets outside the home, if youths are typically members for four or more years, if the group usually meets in the same place, if the group comes from just one part of the neighborhood, if the group has a name, if the group contains older and younger kids, if the respondent meets with the group at least four days a week, and if the respondent takes part in several activities with the group.

8. Spergel (1990) notes that many gang researchers who once believed that gang behavior was not especially serious or lethal (such as Miller and Klein) now have come to the position that such groups are responsible for a large number of homicides and are active participants in widespread narcotics trafficking (see, for example, Miller 1975; Klein and Maxson 1985, 1989).

9. We use the term "fieldwork" in a very broad sense to refer to qualitative study designs that involve some degree of interaction between the researcher and the gang member. They can range from intensive observation conducted by the researcher over extended periods (as in Horowitz 1983, Campbell 1984, Sullivan 1989, Jankowski 1991), through the reports provided by social service workers who deal with particular gangs on a regular basis (as in Short and Strodtbeck 1965, Klein 1971, or Moore 1985), to sets of intensive, unstructured interviews with respondents identified as gang members (as in Vigil 1988 or Hagedorn 1988). Many studies have combined two or more of these techniques into a single research design.

10. In fact, the availability of such information is absolutely essential to the development of processual approaches to gangs.

11. Gang homicides are defined in terms of the criterion used by the Chicago Police Department: "a killing is considered gang-related only if it occurs in the course of an explicitly defined collective encounter between two or more gangs" (Spergel 1984:204). The data reflect the community in which the homicide occurred, not necessarily the residential area of the victim or offender.

12. For a good discussion of media stereotypes of gangs, see Jankowski (1991: Chapter 9).

13. The emphasis on the lower class in these theories reflects the assumption that delinquent behavior was relatively rare in other economic situations or was the result of nonsociological processes such as personal pathology. The impact of the self-report methodology, which provided evidence that apparently contradicted such an assumption, was not yet widespread when these theories were originally formulated.

14. Cohen tends to use the terms "working class" and "lower class" interchangeably in his discussion.

15. Although the neighborhood is noted in passing several times in this work, Cohen does not extensively discuss the local community bases of gang formation. However, given the class-based spatial segregation that characterizes many cities as well as the neighborhood-based schools that were dominant at the time of this work, the extension of his work to local communities per se is straightforward.

16. Contradictory evidence is especially apparent in research that has attempted to document the existence of a subculture of violence in the American South. See Loftin and Parker 1985.

17. Such an orientation is implicit to Spergel's (1984) argument concerning segmented routes to gang activity.

18. In addition to the number of people who have been officially defined as unemployed, this index includes involuntary part-time workers and those discouraged workers who have quit looking for employment.

19. A good sense of the characteristics of the underclass is provided by a recent survey of chronically poor blacks living in eight cities commissioned by the

NAACP (1989; noted in Schaefer and Lamm 1992): women constitute 78 percent of the black underclass, the median income of the chronically poor households during the preceding five years was $4,900, a majority of those had not had a job in the past two years, and over 40 percent had either never held a job or received any employment training.

20. Hagedorn (1988:46), for example, notes that 40 percent of all black high school freshmen in the Milwaukee public schools drop out before their senior year. It should be noted that many of Sullivan's respondents indicated a desire to eventually return to school, or to at least complete a GED. However, very few of them did so in La Barriada.

21. The increasing presence of older gang members also has been noted by Horowitz (1983), Moore (1978), and Hagedorn (1988).

Chapter 6

1. The suspicion arose partly due to the publication of *The Polish Peasant* by Thomas and Znaniecki (1920), which was based on material obtained in the area, the publication of *The Jack-Roller* (1930) by Shaw (the life history of a local resident), and a study that had been conducted by Burgess in a local school. The residents felt that their neighborhood had already been unjustifiably characterized as a hotbed of crime and thought that the presence of the Chicago Area Project might further foster this image.

2. While many of these programs are initiated by agencies outside the local community, some have received widespread media attention and have led to indigenous efforts to develop similar approaches to crime prevention.

3. Housing surveys entail an on-site assessment of the vulnerability of a residence or business to criminal victimization (Lurigio and Rosenbaum 1986).

4. Despite this influence, Addams was fairly ambivalent concerning the religious roots of the movement. See the discussion of Carson (1990).

5. While the settlement workers regarded themselves as serving neighborhoods and not particular individuals, Harold Finestone (1976:124) notes that Clifford Shaw criticized the delinquency prevention efforts of these houses for their treatment of offenders as individuals abstracted from their group and community contexts.

6. One of the most interesting indicators of such condenscension and the vast economic and social gulf that separated the settlers from their neighbors is reflected in recruitment of local residents to serve as personal servants to the staff of Hull House during its early years. See Carson (1990).

7. For a good description of the philosophy and operation of detached worker programs, see Spergel (1964) or Klein (1971).

8. There is some evidence that Shaw may have considered a more confrontational approach but felt constrained because of the representation of the corporate and political interests on the Board of Directors of the CAP. See DuBow et al. (1979:69).

9. It is interesting to note that despite this radical orientation, many of the organizations founded by Alinsky evolved into forms nearly indistinguishable from those of the CAP. See Bennett (1981:217).

10. Therefore, one is not sure how seriously to take Bennett's (1981:178) statement that "as less and less of substantial merit occurred in the CAP [after Shaw's death in 1957], promotional activities took up the slack."

11. What follows is a greatly condensed history of the MFY, drawn extensively from the presentation of Helfgot. Those readers interested in a fuller elaboration of these issues are urged to consult his captivating book.

12. See Helfgot (pp. 80 passim) for a discussion of how the radicalization of the MFY came about.

13. This was a program developed by the U.S. Department of Housing and Urban Development that distributed $40 million to public housing authorities in 39 cities to involve residents in anticrime programs. See U.S. Housing and Urban Development (1980) or Lavrakas (1985).

14. Paul Lavrakas (1985:95) has noted that Congress approved the CACP although previous programs had not provided any real demonstration that such strategies actually led to the decrease of crime.

15. Many programs have been designed to "harden" targets of crime, that is, make them more difficult to victimize. Prominent examples of such activities include permanently marking property with some form of personal identification, taking part in housing surveys that evaluate the susceptibility of a dwelling or business to victimization, and restricting one's activities to particular locations and particular times. While residents are often encouraged by local organizations to engage in these activities and may even be provided information concerning the most effective ways to conduct such hardening, these strategies are essentially individual responses to crime and do not reflect the type of community dynamics upon which we have focused in this book. Therefore, they are not addressed in this chapter.

16. This new orientation has met a significant amount of resistance from some traditionally trained police officers. See Skolnick and Bayley (p. 211) or Sparrow (1988).

17. Skogan, as well as Skolnick and Bayley, note that some of these efforts had a doubtful legal basis.

18. There actually is evidence that the gangs began to negotiate the truce prior to the Los Angeles riots (Sagahuna and Chavez 1992).

References

Akers, Ronald L., Anthony J. La Greca, Christine Sellers, and John Cochran (1987). "Fear of Crime and Victimization among the Elderly in Different Types of Communities." *Criminology* 25: 487–505

Alinsky, Saul D. (1946). *Reveille for Radicals*. Chicago: University of Chicago Press.

Andrews, D.A., Ivan Zinger, Robert D. Hoge, James Bonta, Paul Gendreau, and Francis T. Cullen (1990a). "Does Correctional Treatment Work? A Clinically Relevant and Psychologically Informed Meta-Analysis." *Criminology* 28:369–404.

——— (1990b). "A Human Science Approach or More Punishment and Pessimism: A Rejoinder to Lab and Whitehead." *Criminology* 28:419–430.

Arnold, William R., and Terrance M. Brungardt (1983). *Juvenile Misconduct and Delinquency*. Boston: Houghton Mifflin.

Associated Press (1992). "Gang Violence Continues in Cities." May 19.

Bailey, William C. (1985). "Aggregation and Disaggregation in Cross-Sectional Analyses of Crime Rates: The Case of States, SMSAs and Cities." A paper presented to the Annual Meetings of the American Society of Criminology, San Diego.

Ball-Rokeach, Sandra J. (1973). "Values and Violence: A Test of the Subculture of Violence Thesis." *American Sociological Review* 38:736–749.

——— (1975). "Issues and Non-issues in Testing a Subcultural Thesis: Reply to Magura." *American Sociological Review* 40:836–838.

Bennett, James (1981). *Oral History and Delinquency. The Rhetoric of Criminology*. Chicago: University of Chicago Press.

Bennett, Trevor (1989). "Burglars' Choice of Targets." Pp. 176–192 in *The Geography of Crime*, edited by David J. Evans and David T. Herbert. London: Routledge.

———, and Richard Wright (1984a). *Burglars on Burglary*. Aldershot: Gower.

———(1984b). "Constraints to Burglary: The Offender's Perspective." Pp. 181–200 in *Coping with Burglary*, edited by Ronald V. Clarke and Tim Hope. Boston: Kluwer-Nijhoff.

Berry, Brian J. L., and John D. Kasarda (1977). *Contemporary Urban Ecology*. New York: Macmillan.

Biderman, Albert D., Louise A. Johnson, Jamie McIntyre, and Adrianne W. Weir (1967). *Report on a Pilot Study in the District of Columbia on Victimization and Attitudes toward Law Enforcement*. Washington, DC: U.S. Government Printing Office.

Black, Donald (1970). "The Production of Crime Rates." *American Sociological Review* 35:733–748.

—— (1989). "Social Control as a Dependent Variable." Pp. 1–36 in *Toward A General Theory of Social Control*. Volume 1, edited by Donald Black. Orlando, FL: Academic Press.

——, and M. P. Baumgartner (1980). "On Self Help in Modern Society." Pp. 193–208 in *The Manners and Customs of the Police*, by Donald Black. New York: Academic Press.

Blalock, Hubert M. (1984). "Contextual Effects Models: Theoretical and Methodological Issues." *Annual Review of Sociology* 10:353–372.

Bloch, Herbert A., and Arthur Niederhoffer (1958). *The Gang: A Study in Adolescent Behavior*. New York: Philosophical Library.

Block, Richard (1979). "Community, Environment, and Violent Crime." *Criminology* 17:46–57.

Bluestone, Barry, and Bennett Harrison (1982). *The Deindustrialization of America: Plant Closings, Community Abandonment, and the Dismantling of Basic Industry*. New York: Basic Books.

Boggs, Sarah L. (1965). "Urban Crime Patterns." *American Sociological Review* 30:899–908.

Bogue, Donald J. (1974). *The Basic Writings of Ernest W. Burgess*. Chicago: Community and Family Study Center.

Bookin, Hedy, and Ruth Horowitz (1983). "The End of the Youth Gang: Fad or Fact?" *Criminology* 21:585–602.

Bordua, David J. (1958–1959). "Juvenile Delinquency and 'Anomie': An Attempt at Replication." *Social Problems* 6:230–238.

Bottoms, Anthony E., and Paul Wiles (1986). "Housing Tenure and Residential Crime Careers in Britain." Pp. 101–162 in *Communities and Crime*, edited by Albert J. Reiss, Jr. and Michael Tonry. Chicago: University of Chicago Press.

Bowker, Lee H., Helen Shimata Gross, and Malcolm W. Klein (1980). "Female Participation in Delinquent Gang Activities." *Adolescence* 59:509–519.

Boyd, Lawrence H., Jr., and Gudmund B. Iverson (1979). *Contextual Analysis: Concepts and Statistical Techniques*. Belmont: Wadsworth.

Brantingham, Paul J., and Patricia L. Brantingham (1981). *Environmental Criminology*. Beverly Hills: Sage.

—— (1984). *Patterns in Crime*. New York: Macmillan.

Bureau of the Census (1989) *American Housing Survey for the United States in 1985*. Washington, DC: U.S. Government Printing Office.

Bureau of Justice Statistics (1986). *Crime Prevention Measures*. Washington, DC: U.S. Department of Justice.

—— (1990). *Criminal Victimization in the United States, 1988*. Washington, DC: U.S. Department of Justice.

—— (1991). *Criminal Victimization in the United States, 1989*. Washington, DC: U.S. Department of Justice.

Burgess, Ernest W. (1923). "The Study of the Delinquent as a Person." *American Journal of Sociology* 28:657–680

—— (1925). "The Growth of the City." Pp. 47–62 in *The City*, edited by Robert

E. Park, Ernest W. Burgess, and Roderick D. McKenzie. Chicago: University of Chicago Press.

———, Joseph Lohman, and Clifford R. Shaw (1937). "The Chicago Area Project." Pp. 8–28 in *Yearbook of the National Probation Association*. Reprinted in *The Basic Writings of Ernest W. Burgess*, edited by Donald J. Bogue. Chicago: Community and Family Study Center.

Burnell, James D. (1988). "Crime and Racial Composition in Contiguous Communities as Negative Externalities." *American Journal of Economics and Sociology* 47:177–193.

Bursik, Robert J., Jr. (1984). "Urban Dynamics and Ecological Studies of Delinquency." *Social Forces* 63:393–413.

——— (1986a). "Ecological Stability and the Dynamics of Delinquency." Pp. 35–66 in *Communities and Crime*, edited by Albert J. Reiss, Jr., and Michael Tonry. Chicago: University of Chicago Press.

——— (1986b). "Delinquency Rates as Sources of Ecological Change." Pp. 63–74 in *The Social Ecology of Crime*, edited by James M. Byrne and Robert J. Sampson. New York: Springer-Verlag.

——— (1988). "Social Disorganization and Theories of Crime and Delinquency: Problems and Prospects." *Criminology* 26:519–551.

——— (1989). "Political Decision-Making and Ecological Models of Delinquency: Conflict and Consensus." Pp. 105–117 in *Theoretical Integration in the Study of Deviance and Crime*, edited by Steven F. Messner, Marvin D. Krohn, and Allen E. Liska. Albany: State University of New York Press.

———, and Jim Webb (1982). "Community Change and Patterns of Delinquency." *American Journal of Sociology* 88:24–42.

Buss, Terry F., and F. Stevens Redburn (1983). *Mass Unemployment: Plant Closings and Community Mental Health*. Beverly Hills, CA: Sage.

Byrne, James M., and Robert J. Sampson (1986). "Key Issues in the Social Ecology of Crime." Pp. 1–22 in James M. Byrne and Robert J. Sampson (eds.), *The Social Ecology of Crime*. New York: Springer-Verlag.

Byrne, James M., and Robert J. Sampson (eds.) (1986). *The Social Ecology of Crime*. New York: Springer-Verlag.

Campbell, Ann (1984). *The Girls in the Gang*. Oxford: Basil Blackwell.

Campbell, Donald T. (1969). "Reforms as Experiments." *American Psychologist* 24:409–428.

Cannon, Jane (1990). "Police Make 'Crackdown' Arrests." *Norman Transcript* August 27:1–2.

Cantor, David, and Kenneth C. Land (1985). "Unemployment and Crime Rates in the Post–World War II U.S.: A Theoretical and Empirical Analysis." *American Sociological Review* 50:317–332.

Cantwell, Mary (1990). "Notes for a Casualty in the Battle of the Streets." *New York Times* August 16:C2.

Carson, Mina (1990). *Settlement Folk, Social Thought, and the American Settlement Movement, 1885–1930*. Chicago: University of Chicago Press.

Carter, Ronald L. (1974). *The Criminal's Image of the City*. Unpublished Ph.D. Dissertation. Department of Geography, University of Oklahoma, Norman.

———, and Kim Q. Hill (1978). "Criminal's and Noncriminal's Perceptions of Urban Crime." *Criminology* 16:353–371.

——— (1979). *The Criminal's Image of the City.* New York: Pergamon Press.

Cartwright, Desmond S., and Kenneth J. Howard (1966). "Multivariate Analysis of Gang Delinquency. I. Ecologic Influences." *Multivariate Behavioral Research* 1:321–371.

Castells, Manuel (1983). *The City and the Grassroots: A Cross-Cultural Theory of Urban Social Movements.* Berkeley: University of California Press.

Chambliss, William J. (1973). "The Saints and the Roughnecks." *Society* 11:24–31.

Chamlin, Mitchell B., and Mary B. Kennedy (1991). "The Impact of the Wilson Administration on Economic Crime Rates." *Journal of Quantitative Criminology* 7:357–372.

Chilton, Roland J. (1964). "Continuities in Delinquency Area Research: A Comparison of Studies for Baltimore, Detroit and Indianapolis." *American Sociological Review* 29:71–83.

——— (1987). "Twenty Years of Homicide and Robbery in Chicago: The Impact of the City's Changing Racial and Age Composition." *Journal of Quantitative Criminology* 3:195–214.

Chilton, Roland J., and John P.J. Dussich (1974). "Methodological Issues in Delinquency Research: Some Alternative Analyses of Geographically Distributed Data." *Social Forces* 53:73–80.

Chin, Ko-Lin (1990). "Chinese Gangs and Extortion." Pp. 129–145 in *Gangs in America*, edited by C. Ronald Huff. Newbury Park, CA: Sage.

Chira, Susan (1989). "Programs to Ease Social Problems Burden New York's Poorest Areas." *New York Times* July 16:1+.

Clark, John P., and Eugene P. Wenninger (1962). "Socioeconomic Class and Area Correlates of Illegal Behavior among Juveniles." *American Sociological Review* 27:826–834.

Clarke, Alan H., and Margaret Lewis (1982). "Fear of Crime among the Elderly." *British Journal of Criminology* 22:49–62.

Clarke, Ronald V. (1984). "Opportunity-Based Crime Rates: The Difficulty of Further Refinement." *British Journal of Criminology* 24:74–83.

Clarke, W. A. V., and E. G. Moore (1980). "The Policy Context for Mobility Research." Pp. 1–15 in *Residential Mobility and Public Policy*, edited by W. A. V. Clarke and E. G. Moore. Beverly Hills, CA: Sage.

Cloward, Richard A., and Lloyd Ohlin (1960). *Delinquency and Opportunity.* New York: Free Press.

Cockburn, Alexander (1992). "First a Truce, Now a Real Building Plan." *Los Angeles Times* May 17:M5.

Cohen, Albert K. (1955). *Delinquent Boys.* Glencoe, IL: The Free Press.

Cohen, Bernard (1969). "The Delinquency of Gangs and Spontaneous Groups." Pp. 61–111 in *Delinquency: Selected Studies*, edited by Thorsten Sellin and Marvin E. Wolfgang. New York: Wiley.

——— (1980). *Deviant Street Networks: Prostitution in New York City.* Lexington, MA: Lexington Books.

Cohen, Lawrence E. (1981). "Modeling Crime Trends: A Criminal Opportunity Perspective." *Journal of Research in Crime and Delinquency* 18:138–164.

———, and Marcus Felson (1979). "Social Change and Crime Rate Trends: A Routine Activities Approach." *American Sociological Review* 44:588–608.

———, James R. Kluegel, and Kenneth C. Land (1981). "Social Inequality and Predatory Criminal Victimization: An Exposition and Test of a Formal Theory." *American Sociological Review* 46:505–524.

Conklin, John E. (1975). *The Impact of Crime.* New York: Macmillan.

Cornish, Derek B., and Ronald V. Clarke (1986). *The Reasoning Criminal: Rational Choice Perspectives on Offending.* New York: Springer-Verlag.

Coser, Lewis A. (1982). "The Notion of Control in Sociological Theory." Pp. 13–22 in *Social Control. Views from the Social Sciences*, edited by Jack P. Gibbs, Beverly Hills, CA: Sage.

Costanzo, C. Michael, William C. Halperin, and Nathan Gale (1986). "Criminal Mobility and the Directional Component in Journeys to Crime." Pp. 73–95 in *Metropolitan Crime Patterns*, edited by Robert M. Figlio, Simon Hakim, and George F. Rengert. Monsey, NY: Criminal Justice Press.

Covington, Jeannette, and Ralph B. Taylor (1991). "Fear of Crime in Urban Residential Neighborhoods: Implications of Between- and Within-Neighborhood Sources for Current Models." *Sociological Quarterly* 32:231–249.

Cromwell, Paul F., James N. Olson, and D'Aunn Wester Avary (1991). *Breaking and Entering. An Ethnographic Analysis of Burglary.* Newbury Park, CA: Sage.

Cullen, Francis T., and Karen E. Gilbert (1982). *Reaffirming Rehabilitiation.* Cincinnati, OH: Anderson.

Curry, G. David, and Irving A. Spergel (1988). "Gang Homicide, Delinquency and Community." *Criminology* 26:381–405.

——— (1991). *Youth Gang Involvement and Delinquency. A Report to the National Youth Gang Intervention and Suppression Research and Development Project.* Office of Juvenile Justice and Delinquency Prevention. Washington, DC.

Dawley, David (1992). *A Nation of Lords.* Second Edition. Prospect Heights, IL: Waveland Press.

DiIulio, John J. (1989). "The Impact of Inner-City Crime." *Public Interest* 96:28–46.

Douglas, Jack (1970). *Deviance and Respectability: The Social Construction of Moral Meanings.* New York: Basic Books.

Downes, David M. (1966). *The Delinquent Solution.* New York: Free Press.

DuBow, Fred, Edward McCabe, and Gail Kaplan (1979). *Reactions to Crime: A Critical Review of the Literature.* Washington, DC: U.S. Department of Justice, Law Enforcement Assistance Administration.

DuBow, Fred, and David Emmons (1981). "The Community Hypothesis." Pp. 167–182 in *Reactions to Crime*, edited by Dan A. Lewis. Beverly Hills, CA: Sage.

Eckland-Olson, Sheldon (1989). "Social Control and Relational Disturbance: A Microstructural Paradigm." Pp. 209–233 in *Toward a General Theory of Social Control.* Volume 2, edited by Donald Black. Orlando, FL: Academic Press.

Elliott, Delbert S., David Huizinga, and Suzanne S. Ageton (1985). *Explaining Delinquency and Drug Use.* Beverly Hills, CA: Sage.

Elton, Charles (1927). *Animal Ecology.* New York: Macmillan.

Ennis, Philip H. (1967). *Criminal Victimization in the United States: A Report of a National Survey.* Washington, DC: U.S. Department of Justice.

Erbe, Brigitte Mach (1975). "Race and Socioeconomic Segregation." *American Sociological Review* 40:801–812.

Erlanger, Howard S. (1979). "Estrangement, Machismo, and Gang Violence." *Social Science Quarterly* 60:235–249.

Fagan, Jeffrey (1989). "The Social Organization of Drug Use and Drug Dealing Among Urban Gangs." *Criminology* 27:633–669.

—— (1990). "Social Processes of Delinquency and Drug Use Among Urban Gangs." Pp. 183–219 in *Gangs in America*, edited by C. Ronald Huff. Newbury Park, CA: Sage.

Faris, Robert E. L., and H. Warren Dunham (1939). *Mental Disorders in Urban Areas*. Chicago: University of Chicago Press.

Farley, Reynolds, Suzanne Bianchi, and Dianne Colasanto (1979). "Barriers to the Racial Integration of Neighborhoods: The Detroit Case." *Annals of the American Academy of Political and Social Science* 441:97–113.

Feagin, Joe R., and Robert Parker (1990). *Building American Cities: The Urban Real Estate Game*. Second Edition. Englewood Cliffs, NJ: Prentice-Hall.

Feeney, Floyd (1986). "Robbers as Decision-Makers." Pp. 53–71 in *The Reasoning Criminal* edited by Derek B. Cornish and Ronald V. Clarke. New York: Springer-Verlag.

Felson, Marcus (1986). "Linking Criminal Choices, Routine Activities, Informal Control, and Criminal Outcomes." Pp. 119–128 in *The Reasoning Criminal*, edited by Derek B. Cornish and Ronald V. Clarke. New York: Springer-Verlag.

Felson, Marcus, and Lawrence E. Cohen (1980). "Human Ecology and Crime: A Routine Activity Approach." *Human Ecology* 8:389–406.

Ferraro, Kenneth F., and Randy LaGrange (1987). "The Measurement of Fear of Crime." *Sociological Inquiry* 57:70–101.

Finestone, Harold (1976). *Victims of Change: Juvenile Delinquents in American Society*. Westport, CT: Greenwood Press.

Fischer, Claude S. (1982). *To Dwell among Friends: Personal Networks in Town and City*. Chicago: University of Chicago Press.

Fishman, Gideon, Arye Rattner, and Gabriel Weimann (1987). "The Effect of Ethnicity on Stereotypes." *Criminology* 25:507–524.

Flanagan, Timothy J., and Kathleen Maguire (1990). *Sourcebook of Criminal Justice Statistics—1989*. Washington, DC: U.S. Department of Justice.

Foley, D.L. (1973). "Institutional and Contextual Factors Affecting the Housing Choice of Minority Residents." Pp. 85–147 in *Segregation in Residential Areas*, edited by Amos H. Hawley and Vincent P. Rock. Washington, DC: National Academy of Sciences.

Forer, Lois G. (1990). "Foreword." Pp. xiii–xxv in *Fraternity Gang Rape: Sex, Brotherhood, and Privilege on Campus*, by Peggy Sanday. New York: New York University Press.

Franc, Randy (1986). "Unlikely Allies in a Common War." *Christianity Today* 30(1):54–55.

Fried, Marc (1986). "The Neighborhood in Metropolitan Life: Its Psychosocial Significance." Pp. 331–363 in *Urban Neighborhoods: Research and Policy*, edited by Ralph B. Taylor. New York: Praeger.

Fulwood, Sam, III (1991). "Attitudes on Minorities in Conflict." *Los Angeles Times* January 9:A13.

Furstenberg, Frank F., Jr. (1971). "Public Reaction to Crime in the Streets." *American Scholar* 40:601–610.

Gans, Herbert J. (1962). *The Urban Villagers*. New York: Free Press.

Garafalo, James (1981). "The Fear of Crime: Causes and Consequences." *Journal of Criminal Law and Criminology* 72:839–857.

——, and John Laub. (1979). "The Fear of Crime: Broadening Our Perspectives." *Victimology* 3:242–253.

Garafalo, James (1987). "Reassessing the Lifestyle Model of Criminal Victimization." Pp. 23–42 in *Positive Criminology*, edited by Michael R. Gottfredson and Travis Hirschi. Beverly Hills, CA: Sage.

—— (1990). "The National Crime Survey, 1973–1986: Strengths and Limitations of a Very Large Data Set." Pp. 75–96 in *Measuring Crime*, edited by Doris L. MacKenzie, Phyllis J. Baunach, and Roy R. Roberg. Albany: State University of New York Press.

——, Leslie Seigel, and John Laub (1987). "School-Related Victimizations Among Adolescents: An Analysis of National Crime Survey (NCS) Narratives." *Journal of Quantitative Criminology* 3:321–338.

Garafalo, James, and Maureen McLeed (1989). "The Structure and Operations of Neighborhood Watch Programs in the United States." *Crime and Delinquency* 35:326–344.

Gates, Lauren B., and William M. Rohe (1987). "Fear and Reactions to Crime: A Revised Model." *Urban Affairs Quartlery* 22:425–453.

Gibbs, Jack P. (1989). *Control. Sociology's Central Notion*. Urbana: University of Illinois Press.

Giordano, Peggy C. (1978). "Girls, Guys, and Gangs: The Changing Social Context of Female Delinquency." *Journal of Criminal Law and Criminology* 69:126–132.

Glaser, Daniel (1971). *Social Deviance*. Chicago: Markham.

Glazer, Nathan (1988). *The Limits of Social Policy*. Cambridge, MA: Harvard University Press.

Gold, Martin (1970). *Delinquent Behavior in an American City*. Belmont, CA: Brooks-Cole.

—— (1987). "Social Ecology." Pp. 62–105 in *Handbook of Juvenile Delinquency*, edited by Herbert C. Quay. New York: Wiley.

Gordon, Robert A. (1967). "Issues in the Ecological Study of Delinquency." *American Sociological Review* 32:927–944.

Gottfredson, Michael R. (1981). "On the Etiology of Criminal Victimization." *Journal of Criminal Law and Criminology* 72:712–726.

——, and Travis Hirschi (1990). *A General Theory of Crime*. Stanford: Stanford University Press.

Gottfredson, Stephen D., and Ralph Taylor (1986). "Person-Environment Interactions in the Prediction of Recidivism." Pp. 133–155 in *The Social Ecology of Crime*, edited by James M. Byrne and Robert J. Sampson. New York: Springer-Verlag.

Gould, Leroy (1969). "The Changing Structure of Property Crime in an Affluent Society." *Social Forces* 48:50–59.

Granovetter, Mark S. (1973). "The Strength of Weak Ties." *American Journal of Sociology* 78:1360–1380.

Grasmick, Harold G., and Robert J. Bursik, Jr. (1990). "Conscience, Significant Others, and Rational Choice: Extending the Deterrence Model." *Law and Society Review* 24:837–861.

Greenberg, Stephanie W. (1986). "Fear and Its Relationship to Crime, Neighborhood Deterioration, and Informal Social Control." Pp. 47–62 in *The Social Ecology of Crime*, edited by James M. Byrne and Robert J. Sampson. New York: Springer-Verlag.

Greenberg, Stephanie W., William M. Rohe, and Jay R. Williams (1982a). The "Relationship between Informal Social Control, Neighborhood Crime and Fear: A Synthesis and Assessment of the Research." A paper presented at the annual meeting of the American Society of Criminology, Toronto.

——— (1982b). *Safe and Secure Neighborhoods: Physical Characteristics and Informal Territorial Control in High and Low Crime Neighborhoods.* Washington, DC: National Institute of Justice.

——— (1985). *Informal Citizen Action and Crime Prevention at the Neighborhood Level.* Washington, DC: National Institute of Justice.

Greenberg, Stephanie W., Jay R. Williams, and William M. Rohe (1982c). "Safety in Urban Neighborhoods: A Comparison of Physical Characteristics and Informal Territorial Control in High and Low Crime Neighborhoods." *Population and Environment* 5:141–165.

Guest, Avery M. (1984). "The City." Pp. 277–322 in *Sociological Human Ecology*, edited by Michael Micklin and Harvey M. Choldin. Boulder, CO: Westview Press.

———, and Barrett A. Lee (1987). "Metropolitan Residential Environments and Church Organizational Activities." *Sociological Analysis* 47:335–354.

Hagedorn, John M. (1988). *People and Folks: Gangs, Crime and the Underclass in a Rustbelt city.* Chicago: Lakeview Press.

Hakim, Simon, and George F. Rengert (1981). *Crime Spillover.* Beverly Hills, CA: Sage.

Hallman, Howard W. (1984). *Neighborhoods: Their Place in Urban Life.* Beverly Hills, CA: Sage.

Harré, Rom (1981). "Philosophical Aspects of the Micro-Macro Problem." Pp. 139–160 in *Advances in Social Theory and Methodology. Toward an Integration of Micro- and Macro-Sociologies*, edited by Karin Knorr-Cetina and Aaron V. Cicourel. Boston: Routledge and Kegan Paul.

Harries, Keith D. (1981). "Alternative Denominators in Conventional Crime Rates." Pp. 147–165 in *Environmental Criminology*, edited by Paul J. Brantingham and Patricia L. Brantingham. Beverly Hills, CA: Sage.

Harris, Mary G. (1988). *Las Cholas: Latino Girls and Gangs.* New York: AMS Press.

Hartnagel, Timothy F. (1979). "The Perception and Fear of Crime: Implications for Neighborhood Cohesion, Social Activity, and Community Affect." *Social Forces* 58:176–193.

Haskins, James (1974). *Street Gangs. Yesterday and Today.* New York: Hastings House.

Hawley, Amos H. (1944). "Ecology and Human Ecology." *Social Forces* 23:398–405.

——— (1950). *Human Ecology: A Theory of Urban Structure.* New York: Ronald Press.

Heitgerd, Janet L., and Robert J. Bursik, Jr. (1987). "Extra-Community Dynamics and the Ecology of Delinquency." *American Journal of Sociology* 92:775–787.

Helfgot, Joseph H. (1981). *Professional Reforming. Mobilization for Youth and the Failure of Social Science*. Lexington, MA: Lexington.

Hellman, Daryl A., and Susan Beaton (1986). "The Pattern of Violence in Urban Public Schools: The Influence of School and Community." *Journal of Research in Crime and Delinquency* 23:102–127.

Hershberg, Theodore, Alan N. Burstein, Eugene P. Erickson, Stephanie W. Greenberg, and William L. Yancey (1979). "A Tale of Three Cities: Blacks and Immigrants in Philadelphia: 1850–1880, 1930, and 1970." *Annals of the American Academy of Political and Social Science* 441:55–81.

Hindelang, Michael J., Michael R. Gottfredson, and James Garafalo (1978). *Victims of Personal Crime*. Cambridge: Ballinger.

Hirsch, A.R. (1983). *Making the Second Ghetto: Race and Housing in Chicago 1940–1960*. New York: Cambridge University Press.

Hirschi, Travis (1969). *Causes of Delinquency*. Berkeley: University of California Press.

Hope, Tim, and Mike Hough (1988). "Community Approaches to Reducing Crime." Pp. 1–29 in *Communities and Crime Reduction*, edited by Tim Hope and M. Shaw. London: Her Majesty's Stationery Office.

Horowitz, Ruth (1982). "Adult Delinquent Gangs in a Chicano Community: Masked Intimacy and Marginality." *Urban Life* 11:3–26.

——— (1983). *Honor and the American Dream*. New Brunswick, NJ: Rutgers University Press.

——— (1990). "Sociological Perspectives on Gangs: Conflicting Definitions and Concepts." Pp. 37–54 in *Gangs in America*, edited by C. Ronald Huff. Newbury Park, CA: Sage.

Hough, Mike (1987). "Offender's Choice of Target: Findings from Victim Surveys." *Journal of Quantitative Criminology* 3:355–369.

———, and Helen Lewis (1989). "Counting Crime and Analyzing Risks: The British Crime Survey." Pp. 16–37 in *The Geography of Crime*, edited by David J. Evans and David T. Herbert. London: Routledge.

———, and Pat Mayhew (1983). *The British Crime Survey: First Report*. London: Her Majesty's Stationery Office.

Huckfeldt, R. Robert (1983). "Social Contexts, Social Networks and Urban Neighborhoods: Environmental Constraints on Friendship Choice." *American Journal of Sociology* 89:651–669.

Huff, C. Ronald (1989). "Youth Gangs and Public Policy." *Crime and Delinquency* 35:524–537.

Huizinga, David, Finn-Aage Esbensen, and Anne Wylie Weiher (1991). "Are There Multiple Paths to Delinquency?" *Journal of Criminal Law and Criminology* 82:83–118.

Hunter, Albert J. (1974). *Symbolic Communities*. Chicago: University of Chicago Press.

——— (1978). "Symbols of Incivility: Social Disorder and Fear of Crime in Urban Neighborhoods." A paper presented at the annual meeting of the American Society of Criminology, Dallas.

——— (1985). "Private, Parochial and Public School Orders: The Problem of Crime and Incivility in Urban Communities." Pp. 230–242 in *The Challenge of Social*

Control: Citizenship and Institution Building in Modern Society, edited by Gerald D. Suttles and Mayer N. Zald. Norwood, NJ: Ablex Publishing.

———, and Gerald D. Suttles (1972). "The Expanding Community of Limited Liability." Pp. 44–81 in The Social Construction of Communities, by Gerald D. Suttles. Chicago: University of Chicago Press.

———, and Terry L. Baumer (1982). "Street Traffic, Social Integration and Fear of Crime." Sociological Inquiry 52:122–131.

———, and Suzanne Staggenberg (1988). "Local Communities and Organized Action." Pp. 243–276 in Community Organizations. Studies in Resource Mobilization and Change, edited by Carl Milofsky. New York: Oxford University Press.

Jaehnig, Walter B., David H. Weaver, and Frederick Fico (1981). "Reporting Crime and Fear of Crime in Three Communities." Journal of Communication 31:88–96.

Jankowski, Martin Sanchez (1991). Islands in the Street. Gangs and American Urban Society. Berkeley: University of California Press.

Janowitz, Morris (1951). The Community Press in an Urban Setting. Glencoe, IL: Free Press.

——— (1976). Social Control of the Welfare State. New York: Elsevier.

——— (1978). The Last Half-Century: Societal Change and Politics in America. Chicago: University of Chicago Press.

Jenkins, J. Craig, and Charles Perrow (1977). "Insurgency of the Powerless; Farm Worker Movements (1946–1972)." American Sociological Review 42:249–268.

Jensen, Gary, and David Brownfield (1986). "Gender, Lifestyles, and Victimization: Beyond Routine Activity Theory." Violence and Victims 1:85–89.

Joe, Delbert, and Norman Robinson (1980). "Chinatown's Immigrant Gangs: The New Young Warrior Class." Criminology 18:337–345.

Johnstone, John W. C. (1981). "Youth Gangs and Black Suburbs." Pacific Sociological Review 24:355–375.

Jonassen, Christen T. (1949). "A Re-evaluation and Critique of the Logic and Some Methods of Shaw and McKay." American Sociological Review 14:608–614.

Kapsis, Robert E. (1976). "Continuities in Delinquency and Riot Patterns in Black Residential Areas." Social Problems 23:567–580.

——— (1978). "Residential Succession and Delinquency." Criminology 15:459–486.

Karmen, Andrew (1980). "Race, Inferiority, Crime and Research Taboos." Pp. 81–113 in Taboos in Criminology, edited by Edward Sagarin. Beverly Hills, CA: Sage.

——— (1984). Crime Victims: An Introduction to Victimology. Monterey, CA: Brooks-Cole.

Kasarda, John D., and Morris Janowitz (1974). "Community Attachment in Mass Society." American Sociological Review 39:328–339.

Katzman, M. T. (1980). "The Contribution of Crime to Urban Decline." Urban Studies 17:277–286.

Keiser, R. Lincoln (1969). The Vice Lords: Warriors of the Streets. New York: Holt, Rinehart and Winston.

Kelling, George L. (1987). "Acquiring a Taste for Order: The Community and the Police." *Crime and Delinquency* 33:90–102.

——, and James K. Stewart (1989) *Neighborhoods and Police: The Maintenance of Civil Authority*. Washington, DC: National Institute of Justice.

Kellog, Paul (ed.) (1909–1914). *The Pittsburgh Surveys*. Volumes I–VI. New York: Survey Associates.

Kempf, Kimberly L. (ed.) (1990). *Measurement Issues in Criminology*. New York: Springer-Verlag.

Kempton, Murray (1963). *America Comes of Middle Age*. Boston: Little, Brown.

Kitagawa, Evelyn M., and Karl E. Taeuber (1963). *Local Community Fact Book. Chicago Metropolitan Area*. Chicago: Chicago Community Inventory.

Klein, Malcolm W. (1969). "On Group Context of Delinquency." *Sociology and Social Research* 54:63–71.

——(1971). *Street Gangs and Street Workers*. Englewood Cliffs, NJ: Prentice-Hall.

Klein, Malcolm W., and Cheryl L. Maxson (1985). " 'Rock' Sales in South Los Angeles." *Sociology and Social Research* 69:561–565.

—— (1989). "Street Gang Violence." Pp. 198–234 in *Violent Crime, Violent Criminals*, edited by Neil A. Weiner and Marvin E. Wolfgang. Newbury Park, CA: Sage.

Knight-Ridder News Service (1991). "Philly Old-Timers Learn to Co-exist with Yuppies." March 9.

Kobrin, Solomon (1951). "The Conflict of Values in Delinquency Areas." *American Sociological Review* 16:653–661.

—— (1959). "The Chicago Area Project—A 25 Year Assessment." *Annals of the American Academy of Political and Social Science* 322:20–29.

—— (1971). "The Formal Logical Properties of the Shaw-McKay Delinquency Theory." Pp. 101–131 in *Ecology, Crime and Delinquency*, edited by Harwin L. Voss and D. M. Peterson. New York: Appleton-Century-Crofts.

Kornhauser, Ruth R. (1978). *Social Sources of Delinquency*. Chicago: University of Chicago Press.

Kotlowitz, Alex (1987). "Day to Day Violence Takes a Terrible Toll on Inner City Youth." *Wall Street Journal* CCX(84):1 + .

Krisberg, Barry (1991). "Are You Now or Have You Ever Been a Sociologist?" *Journal of Criminal Law and Criminology* 82:141–155.

Krohn, Harvey, and Leslie W. Kennedy (1985). "Producing Personal Safety: The Effects of Crime Rates, Police Force Size, and Fear of Crime." *Criminology* 23:697–710.

Lab, Steven P., and John T. Whitehead (1990) "From 'Nothing Works' to 'The Appropriate Works': The Latest Stop on the Search for the Secular Grail." *Criminology* 28:405–419.

Lander, Bernard (1954). *Toward an Understanding of Juvenile Delinquency*. New York: Columbia University Press.

Lane, Roger (1980). "Urban Police and Crime in Nineteenth Century American." Pp. 1–45 in *Crime and Justice: A Review of Research*. Volume 2, edited by Norval Morris and Michael Tonry. Chicago: University of Chicago Press.

Laub, John H. (1983). *Criminology in the Making. An Oral History*. Boston: Northeastern University Press.

Lavrakas, Paul J., and Elicia J. Herz (1982). "Citizen Participation in Neighborhood Crime Prevention." *Criminology* 20:479–498.

Lavrakas, Paul J. (1985). "Citizen Self-Help and Neighborhood Crime Prevention Policy." Pp. 87–115 in *American Violence and Public Policy*, edited by Lynn A. Curtis. New Haven, CT: Yale University Press.

Lee, Gary R. (1982). "Sex Differences in Fear of Crime among Older People." *Research and Aging* 4:284–298.

Lerman, Paul (1975). *Community Treatment and Social Control. A Critical Analysis of Juvenile Correctional Policy.* Chicago: University of Chicago Press.

Lewis, Dan A., and Greta Salem (1981). "Community Crime Prevention: An Analysis of a Developing Strategy." *Crime and Delinquency* 27:405–421.

———— (1986). *Fear of Crime, Incivility, and the Production of a Social Problem.* New Brunswick, NJ: Transaction Books.

Lewis, Dan A., J. A. Grant, and Dennis P. Rosenbaum (1985). *The Social Construction of Reform: Prevention and Community Organizations. Final Report to the Ford Foundation.* Volume 2. Evanston, IL: Northwestern University Center for Urban Affairs and Policy Research.

Lichter, S. Robert, Linda S. Lichter, Stanley Rothman, and Daniel Amundson (1987). "Prime-Time Prejudice: TV's Images of Blacks and Hispanics." *Public Opinion* 10: 13–16.

Liska, Allen E., Joseph J. Lawrence, and Andrew Sanchirico (1982). "Fear of Crime as a Social Fact." *Social Forces* 60:760–770.

Liska, Allen E., Andrew Sanchirico, and Mark D. Reed (1988). "Fear of Crime and Constrained Behavior: Specifying and Estimating a Reciprocal Effects Model." *Social Forces* 66:827–837.

Liska, Allen E., and Barbara D. Warner (1991). "Functions of Crime: A Paradoxical Process." *American Journal of Sociology* 96:1441–1463.

Lizotte, Alan J., and David J. Bordua (1980). "Firearms Ownership for Sport and Protection: Two Divergent Models." *American Sociological Review* 45:229–244.

Loeber, Rolf, Magda Stouthamer-Loeber, Welmoet Van Kammen, and David Farrington (1991). "Initiation, Escalation, and Desistance in Juvenile Offending and Their Correlates." *Journal of Criminal Law and Criminology* 82:36–82.

Loftin, Colin, and Robert N. Parker (1985). "The Effect of Poverty on Urban Homicide Rates: An Error in Variables Approach." *Criminology* 23:269–287.

Logan, John R., and Harvey L. Molotch (1987). *Urban Frontiers: The Political Economy of Place.* Berkeley: University of California Press.

Lurigio, Arthur J., and Dennis P. Rosenbaum (1986). "Evaluation Research in Community Crime Prevention: A Critical Look at the Field." Pp. 19–44 in *Community Crime Prevention: Does It Work?* edited by Dennis P. Rosenbaum. Beverly Hills, CA: Sage.

Lynch, James P. (1987). "Routine Activity and Victimization at Work." *Journal of Quantitative Criminology* 3:283–300.

———— (1990). "The Current and Future National Crime Survey." Pp. 97–118 in *Measuring Crime*, edited by Doris L. MacKenzie, Phyllis J. Baunach, and Roy R. Roberg. Albany: State University of New York Press.

MacKenzie, Doris L., Phyllis J. Baunach, and Roy R. Roberg (1990). *Measuring*

Crime: Large-Scale, Long-Range Efforts. Albany: State University of New York Press.

Magura, Stephen (1975). "Is There a Subculture of Violence?" *American Sociological Review* 40:831–836.

Martin, John Bartlow (1944). "A New Attack on Delinquency: How the Chicago Area Project Works." *Harper's Magazine* 188(May):502–512.

Martinson, Robert (1974). "What Works?—Questions and Answers about Prison Reform." *Public Interest* 35:22–54.

——— (1979). "New Findings, New Views: A Note of Caution regarding Sentencing Reform." *Hofstra Law Review* 7:243–258.

Massey, James L., Marvin D. Krohn, and Lisa M. Bonati (1989). "Property Crime and the Routine Activities of Individuals." *Journal of Research in Crime and Delinquency* 26:378–400.

Maxfield, Michael G. (1987a). "Lifestyle and Routine Activity Theories of Crime: Empirical Studies of Victimization, Delinquency, and Offender Decision-Making." *Journal of Quantitative Criminology* 3:275–282.

——— (1987b). "Household Composition, Routine Activity, and Victimization: A Comparative Analysis." *Journal of Quantitative Criminology* 3:301–320.

Maxson, Cheryl. L., Margaret A. Gordon, and Malcolm W. Klein (1985). "Differences Between Gang and Nongang Homicides." *Criminology* 23:209–222.

Maxson, Cheryl L., and Malcolm W. Klein (1990). "Street Gang Violence: Twice as Great or Half as Great?" Pp. 71–100 in *Gangs in America*, edited by C. Ronald Huff. Newbury Park, CA: Sage.

Mayer, Neil S. (1983). "How Neighborhood Development Programs Succeed and Grow: A Survey." Pp. 151–161 in *Neighborhood Policy and Planning*, edited by Philip L. Clay and Robert M. Hollister. Lexington, MA: Lexington.

McAdam, Doug, John D. McCarthy, and Mayer N. Zald (1988). "Social Movements." In *Handbook of Sociology*, edited by Neil J. Smelser. Beverly Hills, CA: Sage.

McBride, Paul W. (1975). *Culture Clash: Immigrants and Reformers, 1880–1920.* San Francisco: R and E Research Associates.

McCarthy, John D., and Mayer N. Zald (1987). "Resource Mobilization and Social Movements: A Partial Theory." Pp. 15–42 in *Social Movements in an Organizational Society*, edited by Mayer N. Zald and John D. McCarthy. New Brunswick, NJ: Transaction.

McCraw, John G. (1986). *The Risk of Criminal Victimization and the Fear of Crime.* Unpublished M.A. thesis, Department of Sociology, University of Oklahoma.

McIver, John P. (1981). "Criminal Mobility: A Review of Empirical Studies." Pp. 20–47 in *Crime Spillover*, edited by Simon Hakim and George F. Rengert. Beverly Hills, CA: Sage.

McKay, Henry D. (1949). "The Neighborhood and Child Conduct." *Annals of the American Academy of Political and Social Science* 262:32–41.

——— (1967). "A Note on Trends in Rates of Delinquency in Certain Areas of Chicago." In *Task Force Report: Juvenile Delinquency and Youth Crime.* President's Commission on Law Enforcement and Administration of Justice. Washington, DC: U.S. Government Printing Office.

Mennel, Robert M. (1973). *Thorns and Thistles. Juvenile Delinquents in the United States, 1825–1940*. Hanover, NH: University Press of New England.

Merry, Sally E. (1981). *Urban Danger. Life in a Neighborhood of Strangers*. Philadelphia: Temple University Press.

Merton, Robert K. (1938). "Social Structure and Anomie." *American Sociological Review* 3:672–682.

Messner, Steven F., and Judith R. Blau (1987). "Routine Leisure Activities: A Macrolevel Analysis." *Social Forces* 65:1035–1052.

Miethe, Terrance, Mark C. Stafford, and J. Scott Long (1987). "Social Differentiation in Criminal Victimization: A Test of Routine Activity/Lifestyle Theories." *American Sociological Review* 52:184–194.

Miller, Alden D., and Lloyd E. Ohlin (1985). *Delinquency and Community. Creating Opportunities and Controls*. Beverly Hills, CA: Sage.

Miller, Walter B. (1958). "Lower Class Culture as a Generating Milieu of Gang Delinquency." *Journal of Social Issues* 14:5–19.

—— (1975). *Violence by Youth Gangs as a Crime Problem in Major American Cities. National Institute for Juvenile Justice and Delinquency Prevention. U.S. Justice Department*. Washington, DC: U.S. Government Printing Office.

—— (1980). "Gangs, Groups, and Serious Youth Crime." Pp. 115–138 in *Critical Issues in Juvenile Delinquency*, edited by David Schichor and Delos H. Kelly. Lexington, MA: D.C. Heath.

Moeller, Gertrude L. (1989). "Fear of Criminal Victimization: The Effect of Neighborhood Racial Composition." *Sociological Inquiry* 59:208–221.

Molotch, Harvey (1976). "The City as a Growth Machine: Toward a Political Economy of Place." *American Journal of Sociology* 82:309–332.

Moore, Joan W. (1978). *Homeboys*. Philadelphia: Temple University Press.

—— (1985). "Isolation and Stigmatization in the Development of an Underclass: The Case of Chicano Gangs in East Los Angeles." *Social Problems* 33:1–12.

—— (1988). "Introduction: Gangs and the Underclass. A Comparative Perspective." Pp. 3–17 in *People and Folks: Gangs, Crime and the Underclass in a Rustbelt City*, by John M. Hagedorn. Chicago: Lake View Press.

Moore, Joan W., Diego Vigil, and Robert Garcia (1983). "Residence and Territoriality in Chicano Gangs." *Social Problems* 31:182–194.

Moorman, Jeanne E. (1980). "Aggregation Bias: An Empirical Demonstration." Pp. 131–156 in *Aggregate Data: Analysis and Interpretation*, edited by Edgar F. Borgatta and David J. Jackson. Beverly Hills, CA: Sage.

Morash, Merry (1983). "Gangs, Groups, and Delinquency." *British Journal of Criminology* 23:309–335.

Morgan, Patricia A. (1978). "The Legislation of Drug Law: Economic Crises and Social Control." *Journal of Drug Issues* 8:53–62.

Morris, Terence (1957). *The Criminal Area*. London: Routledge and Kegan Paul.

Moynihan, Daniel P. (1969). *Maximum Feasible Misunderstanding. Community Action in the War on Poverty*. New York: Free Press.

Muehlbauer, Gene, and Laura Dodder (1983). *The Losers: Gang Delinquency in an American Suburb*. New York: Praeger.

NAACP Legal Defense and Educational Fund (1989). *The Unfinished Agenda on Race in America*. New York: NAACP Legal Defense and Educational Fund.

National Urban League (1991). *Quarterly Economic Report on the African American Worker. First Quarter, 1991. Report No. 28.* June.

Newman, Oscar (1972). *Defensible Space: Crime Prevention through Urban Design.* New York: Macmillan.

Oberschall, Anthony (1973). *Social Conflict and Social Movements.* Englewood Cliffs, NJ: Prentice Hall.

O'Brien, Robert M. (1985). *Crime and Victimization Data.* Beverly Hills, CA: Sage.

O'Connor, Len (1975). *Clout: Mayor Daley and His City.* Chicago: Regnery.

Olson, Mancur (1965). *The Logic of Collective Action. Public Goods and the Theory of Groups.* Cambridge, MA: Harvard University Press.

Pacyga, Dominic A. (1989). "The Russell Square Community Committee: An Ethnic Response to Urban Problems." *Journal of Urban History* 15:155–184.

Park, Robert E. (1926). "The Urban Community as a Special Pattern and a Moral Order." Pp. 3–18 in *The Urban Community*, edited by Ernest W. Burgess. Chicago: University of Chicago Press.

———, and Ernest W. Burgess (1924). *Introduction to the Science of Sociology.* Second Edition. Chicago: University of Chicago Press.

Pate, Anthony, Marlys McPherson, and Glenn Silloway (1987). *The Minneapolis Community Crime Prevention Experiment: Draft Evaluation Report.* Washington, DC: The Police Foundation.

Paternoster, Raymond, Linda E. Saltzman, Gordon P. Waldo, and Theodore Chiricos (1983). "Perceived Risk and Social Control: Do Sanctions Really Deter?" *Law and Society Review* 17:457–479.

Pfohl, Stephen J. (1985). *Images of Deviance and Social Control.* New York: McGraw-Hill.

Platt, Anthony M. (1977). *The Child Savers: The Invention of Delinquency.* Second Edition. Chicago: University of Chicago Press.

Podolefsky, Aaron M. (1983). "Community Response to Crime Prevention: The Mission District." *Journal of Community Action* 1:43–48.

———, and Fred DuBow (1981). *Strategies for Community Crime Prevention.* Springfield, IL: Charles C Thomas.

Pope, Carl E. (1979). "Victimization Rates and Neighborhood Characteristics: Some Preliminary Findings." Pp. 48–57 in *Perspectives on Victimology*, edited by William H. Parsonage. Beverly Hills, CA: Sage.

Portz, John (1990). *The Politics of Plant Closings.* Lawrence: University of Kansas Press.

Prus, Robert, and Styllianoss Irini (1980). *Hookers, Rounders, and Desk Clerks: The Social Organization of the Hotel Community* Salem, WI: Sheffield Publishing.

Rachman, S.J. (1990). *Fear and Courage.* Second Edition. New York: W. H. Freeman.

Rand, Alicia (1986). "Mobility Triangles." Pp. 117–126 in *Metropolitan Crime Patterns*, edited by Robert M. Figlio, Simon Hakim, and George F. Rengert. Monsey, NY: Criminal Justice Press.

——— (1987). "Transitional Life Events and Desistance from Delinquency and Crime." Pp. 134–162 in *From Boy to Man, from Delinquency to Crime*, by Marvin E. Wolfgang, Terence P. Thornberry, and Robert M. Figlio. Chicago: University of Chicago Press.

Reed, John Shelton (1982). *One South. An Ethnic Approach to Regional Culture.* Baton Rouge: Louisiana State University Press.

Reid, Sue Titus (1988). *Crime and Criminology.* Fifth Edition. New York: Holt, Rinehart and Winston.

Reinarman, Craig, and Harry G. Levine (1989). "The Crack Attack: Politics and Media in America's Latest Drug Scare." Pp. 115–137 in *Images of Issues: Typifying Contemporary Social Problesm,* edited by Joel Best. New York: Aldine de Gruyter.

Reiss, Albert J., Jr. (1967). *Studies in Crime and Law Enforcement in Major Metropolitan Areas.* Washington, DC: U.S. Government Printing Office.

—— (1988). "Co-Offending and Criminal Careers." Pp. 117–170 in *Crime and Justice: A Review of Research.* Vol. 10, edited by Michael Tonry and Norval Morris. Chicago: University of Chicago Press.

Rengert, George F., and John Wasilchick (1985). *Suburban Burglary: A Time and Place for Everything.* Springfield, IL: Charles C Thomas.

Rengert, George F. (1989). "Behavioral Geography and Criminal Behavior." Pp. 161–175 in *The Geography of Crime,* edited by David J. Evans and David T. Herbert. London: Routledge.

Reppeto, Thomas A. (1974). *Residential Crime.* Cambridge, MA: Ballinger.

Riggs, David S., and Dean G. Kilpatrick (1990). "Families and Friends: Indirect Victimization by Crime." Pp. 120–138 in *Victims of Crime: Problems, Policies and Programs,* edited by Arthur J. Lurigio, Wesley G. Skogan, and Robert C. Davis. Beverly Hills, CA: Sage.

Roncek, Dennis W. (1981). "Dangerous Places: Crime and Residential Environment." *Social Forces* 60:74–96.

—— (1987a). "Changing Crime Patterns in Two Major Cities: The Cases of Cleveland and San Diego." A paper presented at the annual meeting of the American Society of Criminology, Montreal.

—— (1987b). "Racial Composition and Crime: A Comparative Analysis of Intraurban Relationships." A paper presented at the annual meeting of the American Society of Criminology, Montreal.

Rosen, L., and S. H. Turner (1967). "An Evaluation of the Lander Approach to the Ecology of Crime." *Social Problems* 15:189–200.

Rosenbaum, Dennis P. (1986). "The Problem of Crime Control." Pp. 11–18 in *Community Crime Prevention: Does It Work?* edited by Dennis P. Rosenbaum. Beverly Hills, CA: Sage.

—— (1987). "The Theory and Research behind Neighborhood Watch: Is It a Sound Fear and Crime Reduction Strategy?" *Crime and Delinquency* 33:103–134.

Ross, Edward A. (1901). *Social Control: A Survey of the Foundations of Order.* New York: Macmillan.

Rothman, Jack (1979). "Three Models of Community Organization Practice: Their Mixing and Phasing." Pp. 25–45 in *Strategies in Community Organization,* edited by F. M. Cox. Itasca, IL: Peacock.

Royko, Mike (1971). *Boss: Richard J. Daley of Chicago.* New York: Dutton

Russell Sage Foundation (1914). *West Side Studies.* New York: Survey Associates.

Sagahuna, Louis, and Stephanie Chavez (1992). "8-Trey Crips Have Chilling Crime Record." *Los Angeles Times* May 13:A1, 20.

Sampson, Robert J. (1983). "Structural Density and Criminal Victimization." *Criminology* 21:276–293.

—— (1985). "Neighborhood and Crime: The Structural Determinants of Personal Victimization." *Journal of Research in Crime and Delinquency* 22:7–40.

—— (1986). "Neighborhood Family Structure and the Risk of Personal Victimization." Pp. 25–46 in *The Social Ecology of Crime*, edited by James M. Byrne and Robert J. Sampson. New York: Springer-Verlag.

—— (1987a). "Communities and Crime." Pp. 91–114 in *Positive Criminology*, edited by Michael R. Gottfredson and Travis Hirschi. Beverly Hills, CA: Sage.

—— (1987b). "Does an Intact Family Reduce Burglary Risk for Its Neighbors?" *Sociology and Social Research* 71:204–207.

—— (1988). "Local Friendship Ties and Community Attachment in Mass Society: A Multilevel Systemic Model." *American Sociological Review* 53:766–779.

—— (1989). "The Promises and Pitfalls of Macro-Level Research." *The Criminologist* 14:1+.

——, and Thomas C. Castellano (1982). "Economic Inequality and Personal Victimization." *British Journal of Criminology* 22:363–385.

——, and W. Byron Groves (1989). "Community Structure and Crime: Testing Social-Disorganization Theory." *American Journal of Sociology* 94:774–802.

——, and Janet L. Lauritsen (1990). "Deviant Lifestyles, Proximity to Crime, and the Offender-Victim Link in Personal Violence." *Journal of Research in Crime and Delinquency* 27:110–139.

——, and John D. Wooldredge (1987). "Liking the Micro- and Macro-Level Dimensions of Lifestyle-Routine Activity-Opportunity Models of Predatory Victimization." *Journal of Quantitative Criminology* 3:371–393.

Sanday, Peggy (1990). *Fraternity Gang Rape: Sex, Brotherhood, and Privilege on Campus.* New York: New York University Press.

Sanders, Marion K. (1970). *The Professional Radical: Conversations with Saul Alinsky.* New York: Harper and Row.

Savitz, Leonard D., Lawrence Rosen, and Michael Lalli (1980). "Delinquency and Gang Membership as Related to Victimization." *Victimology* 5:152–160.

Schaefer, Richard T., and Robert P. Lamm (1992). *Sociology.* Fourth Edition. New York: McGraw-Hill.

Schlossman, Steven, and Michael Sedlak (1983a). *The Chicago Area Project Revisited.* Santa Monica, CA: Rand Corporation.

—— (1983b). "The Chicago Area Project Revisited." *Crime and Delinquency* 29:398–462.

Schlossman, Steven, Gail Zellman, Richard Skavelson, Michael Sedlak, and Jan Cobb (1984). *Delinquency Prevention in South Chicago. A Fifty Year Assessment of the Chicago Area Project.* Santa Monica, CA: Rand Corporation.

Schmid, Calvin F. (1960a). "Urban Areas: Part I." *American Sociological Review* 25:527–542.

—— (1960b). "Urban Areas: Part II." *American Sociological Review* 25:655–678.

Schrager, Laura S., and James F. Short, Jr. (1980). "How Serious a Crime? Perceptions of Organizational and Common Crimes." Pp. 14–31 in *White Collar Crime: Theory and Research*, edited by Gilbert Geis and Ezra Stotland. Beverly Hills, CA: Sage.

Schuerman, Leo A., and Solomon Kobrin (1983a). Community Careers in Crime. Unpublished Manuscript.

—— (1983b). "Crime and Urban Ecological Processes: Implications for Public Policy. A paper presented at the annual meeting of the American Society of Criminology, Denver.

—— (1986). "Community Careers in Crime." Pp. 67–100 in *Communities and Crime*, edited by Albert J. Reiss, Jr., and Michael Tonry. Chicago: University of Chicago Press.

Schuman, Howard, Charlotte Steel, and Lawrence Bobo (1985). *Racial Attitudes in America*. Cambridge, MA: Harvard University Press.

Schwab, William A. (1987). "The Predictive Value of Three Ecological Models: A Test of the Life-Cycle, Arbitrage, and Composition Models of Neighborhood Change." *Urban Affairs Quarterly* 23:295–308.

Schwartz, Gary (1987). *Beyond Confromity or Rebellion*. Chicago: University of Chicago Press.

Sechrest, Lee B., Susan O. White, and Elizabeth D. Brown (eds.) (1979). *The Rehabilitation of Criminal Offenders: Problems and Prospects*. Washington, DC: National Academy Press.

Sellin, Thorsten, and Marvin E. Wolfgang (1964). *The Measurement of Delinquency*. New York: Wiley.

Shannon, Lyle W. (1982). "The Relationship of Juvenile Delinquency and Adult Crime to the Changing Ecological Structure of the City." Executive Report submitted to the National Institute of Justice.

—— (1988). *Criminal Career Continuity. Its Social Context*. New York: Human Sciences Press.

—— (1991). *Changing Patterns of Delinquency and Crime. A Longitudinal Study in Racine*. Bouder, CO: Westview Press.

Shaw, Clifford R., Frederick M. Zorbaugh, Henry D. McKay, and Leonard S. Cottrell (1929). *Delinquency Areas*. Chicago: University of Chicago Press.

Shaw, Clifford R. (1930). *The Jack-Roller: A Delinquent Boy's Own Story*. Chicago: University of Chicago Press.

—— (1931). *the Natural History of a Delinquent Career*. Chicago: University of Chicago Press.

——, Henry D. McKay, and James F. McDonald (1938). *Brothers in Crime*. Chicago: University of Chicago Press.

Shaw, Clifford R., and Henry D. McKay (1931). *Social Factors in Juvenile Delinquency*. National Commission on Law Observation and Enforcement, No. 13, Report on the Causes of Crime, Volume II. Washington, DC: U.S. Government Printing Office.

—— (1942). *Juvenile Delinquency and Urban Areas*. Chicago: University of Chicago Press.

—— (1949). "Rejoinder." *American Sociological Review* 14:614–617.

—— (1969). *Juvenile Delinquency and Urban Areas*. Second Edition. Chicago: University of Chicago Press.

Sheley, Joseph F. (1991). "Conflict and Criminal Law." Pp. 21–39 in *Criminology: A Contemporary Handbook*, edited by Joseph F. Sheley. Belmont, CA: Wadsworth.

Shevky, Eshref, and Wendell Bell (1955). *Social Area Analysis*. Stanford, CA: Stanford University Press.

Sherman, Lawrence W. (1986). "Policing Communities: What Works?" Pp. 343–386 in *Communities and Crime*, edited by Albert J. Reiss, Jr., and Michael Tonry. Chicago: University of Chicago Press.

———, Patrick R. Gartin, and Michael E. Buerger (1989). "Hot Spots of Predatory Crime: Routine Activities and the Criminology of Place." *Criminology* 27:27–56.

Short, James F., Jr. (1963). "Introduction to the Abridged Edition." Pp. xv–liii in *The Gang*, by Frederic Thrasher. Chicago: University of Chicago Press.

——— (1969). "Introduction to the Revised Edition." Pp. xxv–liv in *Juvenile Delinquency and Urban Areas*. Second Edition, Clifford R. Shaw and Henry D. McKay. Chicago: University of Chicago Press.

——— (1990). *Delinquency and Society*. Englewood Cliffs, NJ: Prentice-Hall.

Short, James F., Jr., and Fred L. Strodtbeck (1965). *Group Process and Gang Delinquency*. Chicago: University of Chicago Press.

——— (1974). "Preface, 1974." Pp. v–xiv in *Group Process and Gang Delinquency*. Second Printing, by James F. Short, Jr., and Fred L. Strodtbeck. Chicago: University of Chicago Press.

Siegel, Larry J. (1986). *Criminology*. Second Edition. St. Paul, MN: West.

Simcha-Fagan, Ora, and Joseph E. Schwartz (1986). "Neighborhood and Delinquency: An Assessment of Contextual Effects." *Criminology* 24:667–703.

Singer, Simon I (1987). "Victims in a Birth Cohort." Pp. 163–179 in *From Boy to Man, from Delinquency to Crime*, edited by Marvin E. Wolfgang, Terence P. Thornberry, and Robert M. Figlio. Chicago: University of Chicago Press.

Skogan, Wesley G., and Michael G. Maxfield (1981). *Coping with Crime: Individual and Neighborhood Reactions*. Beverly Hills, CA: Sage.

Skogan, Wesley G. (1986). "Fear of Crime and Neighborhood Change." Pp. 203–229 in *Communities and Crime*, edited by Albert J. Reiss, Jr., and Michael Tonry. Chicago: University of Chicago Press.

——— (1988). "Community Organizations and Crime." Pp. 39–78 in *Crime and Justice: A Review of Research*. Volume 10, edited by Michael Tonry and Norval Morris. Chicago: University of Chicago Press.

——— (1990). *Disorder and Decline: Crime and the Spiral of Decay in American Neighborhoods*. New York: Free Press.

Skolnick, Jerome H., and David H. Bayley (1986). *The New Blue Line: Police Innovation in Six American Cities*. New York: Free Press.

Slovak, Jeffrey S. (1986). "Attachments in the Nested Community: Evidence From a Case Study." *Urban Affairs Quarterly* 21:575–597.

——— (1987). "Police Organiztion and Policing Environment: Case Study of a Disjuncture." *Sociological Focus* 20:77–94.

Smelser, Neil J. (1962). *Theory of Collective Behavior*. New York: Free Press.

Smith, Douglas A. (1986). "The Neighborhood Context of Police Behavior." Pp. 313–341 in *Communities and Crime*, edited by Albert J. Reiss, Jr., and Michael Tonry. Chicago: University of Chicago Press.

Smith, Douglas A., and G. Roger Jarjoura (1988). "Social Structure and Criminal Victimization." *Journal of Research in Crime and Delinquency* 25:27–52.

—— (1989). "Household Characteristics, Neighborhood Composition, and Victimization Risk." *Social Forces* 68:621–640.

Smith, Joel (1991). "A Methodology for Twenty-First Century Sociology." *Social Forces* 70:1–17.

Smith, Susan J. (1986). *Crime, Space and Society*. Cambridge: Cambridge University Press.

—— (1989). "Social Relations, Neighborhood Structure, and the Fear of Crime in Britain." Pp. 193–227 in *The Geography of Crime*, edited by David J. Evans and David Herbert. London: Routledge.

Snodgrass, Jon (1976). "Clifford R. Shaw and Henry D. McKay: Chicago Criminologists." *British Journal of Criminology* 16:1–19.

Sonleitner, Nancy, Peter Wood, and Harold G. Garsmick (1992). "Prejudice, Stereotype, and the Contact Hypothesis." A paper presented at the annual meetings of the Southwest Sociological Society, Austin, Texas.

Sorrentino, Anthony (1959). "The Chicago Area Project after 25 Years." *Federal Probation* 23:40–45.

Sparrow, Malcolm K. (1988). *Implementing Community Policing*. Washington DC: National Institute of Justice.

Spergel, Irving A. (1964). *Street Gang Work: Theory and Practice*. Reading, MA: Addison-Wesley.

—— (1984). "Violent Gangs in Chicago: In Search of Social Policy." *Social Service Review* 58:199–226.

—— (1986). "Violent Gangs in Chicago: A Local Community Approach." *Social Service Review* 60:94–131.

—— (1990). "Youth Gangs: Continuity and Change." Pp. 171–275 in *Crime and Justice: A Review of Research*. Volume 12, edited by Michael Tonry and Norval Morris. Chicago: University of Chicago Press.

Spergel, Irving A., and John Korbelik (1979). "The Local Community Service System and ISOS: An Interorganizational Analysis." Executive Report to the Illinois Law Enforcement Commission.

Spergel, Irving A., and G. David Curry (1988). "Socialization to Gangs: Preliminary Baseline Report." School of Social Service Administration, University of Chicago.

—— (1990). "Strategies and Perceived Agency Effectiveness in Dealing with the Youth Gang Problem." Pp. 288–317 in *Gangs in America*, edited by C. Ronald Huff. Newbury Park, CA: Sage.

Stafford, Mark C., and Omer R. Galle (1984). "Victimization Rates, Exposure to Risk, and Fear of Crime." *Criminology* 22:173–185.

Stark, Rodney (1987). "Deviant Place: A Theory of the Ecology of Crime." *Criminology* 25:893–909.

Stein, Ben (1985). " 'Miami Vice': It's So Hip You'll Want to Kill Yourself." *Public Opinion* 8:41–43.

Stinchcombe, Arthur L., Carol Heimer, R. A. Iliff, Kimberly Scheppele, Thomas W. Smith, and D. Garth Taylor (1978). *Crime and Punishment in Public Opinion: 1948–1974*. Chicago: National Opinion Research Center.

Stinchcombe, Arthur L., Rebecca Adams, Carol Heimer, Kimberly Scheppele, Thomas W. Smith, and D. Garth Taylor (1980). *Crime and Punishment in Public Opinion*. San Francisco: Jossey-Bass.

Sullivan, Mercer L. (1989). *Getting Paid: Youth Crime and Work in the Inner City*. Ithaca, NY: Cornell University Press.

Sumner, William Graham (1906, 1959). *Folkways*. New York: Dover.

Sutherland, Edwin H. (1934). *Principles of Criminology*. Second Edition. Philadelphia: J. B. Lippincott.

Suttles, Gerald D. (1968). *The Social Order of the Slum*. Chicago: University of Chicago Press.

——— (1972). *The Social Construction of Communities*. Chicago: University of Chicago Press.

Swigert, Victoria L., and Ronald A. Farrell (1976). *Murder, Inequality, and the Law*. Lexington, MA: Lexington Books.

Taub, Richard D., George P. Surgeon, and Sara Lindholm (1977). "Urban Voluntary Association, Locality-Based and Externally Induced." *American Journal of Sociology* 83:425–442.

Taub, Richard P., D. Garth Taylor, and Jan D. Dunham (1981). "Neighborhoods and Safety." Pp. 103–119 in *Reactions to Crime*, edited by Dan A. Lewis. Beverly Hills, CA: Sage.

——— (1984). *Paths of Neighborhood Change: Race and Crime in Urban America*. Chicago: University of Chicago Press.

Taylor, D. Garth, Richard P. Taub, and Bruce L. Peterson (1986). "Crime, Community Organization, and Causes of Neighborhood Decline." Pp. 161–177 in *Metropolitan Crime Patterns*, edited by Robert M. Figlio, Simon Hakim, and George F. Rengert. Monsey, NY: Criminal Justice Press.

Taylor, Ralph B., S.A. Schumaker, and Stephen D. Gottfredson (1985). "Neighborhood-Level Link between Physical Features and Local Sentiments: Deterioration, Fear of Crime, and Confidence." *Journal of Architectural Planning and Research* 2:261–275.

Taylor, Ralph B., and Margaret Hale (1986). "Testing Alternative Models of Fear of Crime." *Journal of Criminal Law and Criminology* 77:151–189.

Taylor, Ralph B., and Jeanette Covington (1988). "Neighborhood Changes in Ecology and Violence." *Criminology* 26:553–589.

Terry, Don (1992). "Hope and Fear in Los Angeles as Deadly Gangs Call Truce." *New York Times* May 12:A1, 11.

Thomas, William I., and Florian Znaniecki (1920). *The Polish Peasant in Europe and America*. Volume IV. Boston: Gorham Press.

Thornberry, Terence P., Alan J. Lizotte, Marvin D. Krohn, Margaret Farnsworth, and Sung Joon Jang. (1991). "Testing Interactional Theory: An Examination of Reciprocal Causal Relationships among Family, School, and Delinquency." *Journal of Criminal Law and Criminology* 82:3–35.

Thrasher, Frederic M. (1927). *The Gang*. Chicago: University of Chicago Press.

Tittle, Charles R. (1989). "Influences on Urbanism: A Test of Three Perspectives. *Social Problems* 36:270–288.

Tracy, Paul E., Marvin E. Wolfgang, and Robert M. Figlio (1990). *Delinquency Careers in Two Birth Cohorts*. New York: Plenum Press.

Taun, Yi-Fu (1979). *Landscapes of Fear*. New York: Pantheon.

Turner, A. G. (1972). *The San Jose Methods Test of Known Crime Victims*. National Criminal Justice Information and Statistics Service, Law Enforcement Assistance Administration. Washington, DC: USGPO.

Tyler, Tom R. (1984). "Assessing the Risk of Crime Victimization: The Integration of Personal Victimization Experiences and Socially Transmitted Information." *Journal of Social Issues* 40:27–38.

———, and Fay Lomax Cook (1984). "The Mass Media and Judgments of Risk: Distinguishing Impact on Personal and Societal Level Judgements." *Journal of Personality and Social Psychology* 47:693–708.

United States Department of Housing and Urban Development (1980). *Urban Initiatives Anti-Crime Program. First Annual Report to Congress.* Washington, DC: U.S. Government Printing Office.

United States General Accounting Office (1989). *Nontraditional Organized Crime.* Washington, DC: U.S. Government Printing Office.

Unnever, James D. (1987). Review of *The Social Ecology of Crime*, edited by James M. Byrne and Robert J. Sampson. *Contemporary Sociology* 16:845–846.

van der Wurff, Adri, Leendert van Staaldvinem, and Peter Stronger (1989). "Fear of Crime in Residential Environments: Testing a Social Psychological Model." *Journal of Social Psychology* 129:141–160.

Vigil, James D. (1983). "Chicano Gangs: One Response to Mexican Urban Adaptation." *Urban Anthropology* 12:45–75.

——— (1988). *Barrio Gangs: Street Life and Identity in Southern California.* Austin: University of Texas Press.

———, and John M. Long (1990). "Emic and Etic Perspectives on Gang Culture: The Chicano Case." Pp. 55–68 in *Gangs in America*, edited by C. Ronald Huff. Newbury Park, CA: Sage.

Vigil, James D., and Steve Chong Yun (1990). "Vietnamese Youth Gangs in Southern California." Pp. 146–162 in *Gangs in America*, edited by C. Ronald Huff. Newbury Park, CA: Sage.

Wacquant, Loic J. D., and William J. Wilson (1989). "The Cost of Racial and Class Exclusion in the Inner City." *Annals of the American Academy of Political and Social Science* 501:8–25.

Walker, Samuel (1985). *Sense and Nonsense about Crime. A Policy Guide.* Monterey, CA: Brooks/Cole.

Wallace, Linda S. (1991). "Big-City Terror Stalks Small-Town America." *Knight-Ridder Newspaper Service* December 26.

Walsh, Dermot (1986). "Victim Selection Procedures among Economic Criminals: The Rational Choice Perspective." Pp. 39–52 in *The Reasoning Criminal*, edited by Derek B. Cornish and Ronald V. Clarke. New York: Springer-Verlag.

Ward, David (1989). *Poverty, Ethnicity, and the American City, 1840–1925. Changing Conceptions of the Slum and the Ghetto.* Cambridge: Cambridge University Press.

Warr, Mark (1984). "Fear of Victimization: Why Are Women and the Elderly More Afraid?" *Social Science Quarterly* 65:681–702.

——— (1985). "Fear of Rape among Urban Women." *Social Problems* 32:238–250.

——— (1990). "Dangerous Situations: Social Context and Fear of Criminal Victimization." *Social Forces* 68:891–907.

——— (1991). "America's Perception of Crime and Punishment." Pp. 5–19 in *Criminology: A Contemporary Handbook*, edited by Joseph F. Sheley. Belmont, CA: Wadsworth.

————, and Mark C. Stafford (1983). "Fear of Victimization: A Look at the Proximate Causes. " *Social Forces* 61:1033–1043.

Weicher, J. C. (1980). *Housing: Federal Policies and Programs*. Washington, DC: American Enterprise Institute.

Weissman, Harold H. (1969). "Overview of the Community Development Program." Pp. 23–28 in *Community Development in the Mobilization for Youth Experience*, edited by Harold H. Weissman. New York: Association Press.

Whyte, William F. (1981). *Street Corner Society*. Third Edition. Chicago: University of Chicago Press.

Will, Jeffry A. (1991). "Crime, Neighborhood Perceptions, and the Underclass: An Empirical Tip Toe through the Theory." A paper presented to the annual meeting of the American Society of Criminology, San Francisco.

Williams, Kirk, and Richard Hawkins (1989). "The Meaning of Arrest for Wife Assault." *Criminology* 27:163–181.

Williams, P. A. (1988). "A Recursive Model of Intraurban Trip-making." *Environment and Planning A* 20:535–546.

Wilson, James Q., and Richard Herrnstein (1985). *Crime and Human Nature*. New York: Simon and Schuster.

Wilson, James Q., and George L. Kelling (1982). "Broken Windows." *Atlantic Monthly* March:29–38.

Wilson, William J. (1980). *The Declining Significance of Race: Blacks and Changing American Institutions*, Second Edition. Chicago: University of Chicago Press.

———— (1987). *The Truly Disadvantaged*. Chicago: University of Chicago Press.

———— (1988). "The Ghetto Underclass and the Social Transformation of the Inner City." *Black Scholar* 19:10–17.

———— (1991). "Studying Inner-City Social Dislocations: The Challenge of Public Agenda Research." *American Sociological Review* 56:1–14.

Wirth, Louis (1938). "Urbanism as a Way of Life." *American Journal of Sociology* 44:1–24.

————, and Margaret Furez (1938). *Local Community Fact Book*. Chicago: Chicago Recreation Commission.

Wolfgang, Marvin E., and Franco Ferracuti (1982). *The Subculture of Violence: Towards an Integrated Theory in Criminology*. Beverly Hills, CA: Sage.

Wolfgang, Marvin E., Robert M. Figlio, Paul E. Tracy, and Simon I. Singer (1985). *The National Survey of Crime Severity*. Washington, DC: U.S. Department of Justice.

Zatz, Marjorie S. (1987). "Chicano Youth Gangs and Crime: The Creation of a Moral Panic. *Contemporary Crises* 11:129–158.

Zinsmeister, Karl (1990). "Growing up Scared." The *Atlantic Monthly* June:49–66.

Acknowledgements

R eaders of academic and policy literature routinely glance quickly through the acknowledgments section of books without giving a great deal of thought to the intellectual and emotional debts that are reflected in those few pages. We happily anticipated the writing of this part of the book, for it would signify the celebratory moment when a long process of sleep deprivation and caffeine abuse finally had come to an end. Now, however, as we actually compose our acknowledgments, our mood is one of deep humility rather than festivity as we come face to face with the enormous personal debts that we have accumulated and realize that a simple acknowledgment cannot begin to do justice to the support we have received. It is a terrifying feeling to realize that no matter how carefully we craft this section, we are bound to omit someone who has contributed a great deal to the creation of this book.

The most difficult task is to provide suitable recognition to those who have had the most influence on the intellectual development of the book, for many of the roots of our argument first began to emerge during our days in graduate school at the Universities of Chicago and North Carolina. Thus, although they had little to do with this work per se, the influence of such sociologists as Donald Bogue, Jim Coleman, Amos Hawley, Morris Janowitz, John Kasarda, Barry Schwartz, and Gerry Suttles certainly is apparent. Likewise, many members of the general criminology community consistently have been supportive and constructively critical of our work in the area of community and crime. Especially deserving of our gratitude have been Jim Byrne, David Farrington, Joel Garner, Travis Hirschi, John Laub, Al Liska, Jim Lynch, Steve Messner, Winnie Reed, Al Reiss, Rob Sampson, Jim Short, Neil Shover, Ora Simcha-Fagan, Sally Simpson, Wes Skogan, Doug Smith, Irv Spergel, Ralph Taylor, and Charles Tittle. Special places in this pantheon are reserved for Sol Kobrin and Lyle Shannon, who have devoted their careers to studying many of the issues raised in this book and have persisted even when the "wise" career choice may have been to move to other substantive topics.

Without question, our orientation in this book owes the greatest intellectual debt to the years that Bursik worked at the Institute for Juvenile Research in Chicago. The institute was an exciting arena of stimulating exchange during that period, and the full influence of Don Merten, Gary

Schwartz, Joe Puntil, and (especially) Jim Webb on the development of our ideas is impossible to specify in words. That debt will never be paid.

Of course, there are more immediate influences that led to the publication of this book. George Ritzer first approached us to see if we would be interested in writing a book for inclusion in a series that he was editing for Lexington Books. He consistently has been supportive, encouraging, and understanding during this endeavor. Likewise, Beth Anderson of Lexington Books never has failed to provide as much assistance as we could possibly desire. Perhaps we have simply been lucky, but we encountered absolutely none of the horror stories concerning the publication of a book that commonly get circulated through the grapevine.

Our colleagues in the Department of Sociology of the University of Oklahoma provided us invaluable support and assistance during this enterprise and always offered very insightful (and prompt!) comments on chapter drafts that had been circulated. We would like to especially note the contributions of Bruce Arneklev, Brenda Blackwell, Mitch Chamlin, John Cochran, Jennifer Hackney, Craig St. John, and Jeff Will.

Although the contributions of the aforementioned people are incalculable, they were exposed to this enterprise only during working hours. A very special group of family and friends consented to be exposed to this effort nearly around the clock. In retrospect, we must have seemed to have all the desirable personal qualities of slugs at times during the writing; nevertheless, they very patiently tolerated missed dinners and poker games as well as periods of testiness. The most consistently abused friends easily were Dennis Bean, Jaye Bryan, Ron Chance, Randy Fannon, Vickie and Charlie Farrell, and Tim Morgan. Jaye, Ron, and Tim, in particular, showed the patience of Trappist monks. Bursik promises never to eat the chicken salad again without asking.

As always, the best is saved for last. This book simply could not have been written without the efforts of Jennifer Bursik and Mary Grasmick, who easily could be considered to be the third and fourth authors of the book. It takes a special kind of person to be willing to be married to an academic, and that willingness got us through the tough parts of this process. Jennifer deserves a special commendation for reading every draft of every chapter several times, offering suggestions concerning style and content drawn from her experience as a newspaper reporter. It was a tough education, but we learned our lesson concerning split infinitives. Nevertheless, Jennifer, "operationalization" is a perfectly legitimate sociological word.

Finally, this book is dedicated to our sons, Travis Bursik and Jake Grasmick. We wrote this book in the hope that it might make at least a little difference in the future, and that desire was with both of you in mind. May the world never get to the point that exploring neighborhoods cannot provide you with the same excitement, sense of wonder, and education that it has provided us.

Index

About the Authors

Robert J. Bursik, Jr. received his Ph.D. in Sociology from the University of Chicago in 1980 and is currently an Associate Professor and Chair of Sociology and Co-Director of the Center for the Study of Crime, Delinquency and Social Control at the University of Oklahoma. His body of research in the area of neighborhoods and crime has attempted to integrate the traditional social disorganization model into the larger contexts of a systemic theory of community organization and, more recently, the changing political economies of urban areas. He has served on the editorial boards of the *American Journal of Sociology, Social Forces, Criminology,* and the *Journal of Research in Crime and Delinquency,* and has been a member of the Executive Board of the American Society of Criminology.

Harold G. Grasmick received his Ph.D. in Sociology in 1973 from the University of North Carolina and is currently a Professor of Sociology and Co-Director of the Center for the Study of Crime, Delinquency and Social Control at the University of Oklahoma. In addition to his work in the area of social disorganization, he has written extensively concerning the links between formal and informal sanctions and self-imposed internal sanctions. He has served on the editorial boards of *Law and Policy,* the *Journal of Criminal Law and Criminology,* and *Criminology,* and for five years was a member of the National Research Council panel that awards National Science Foundation Graduate Minority Fellowships.

Made in the USA
Monee, IL
11 January 2023

25081204R00134